Geography

GLOBAL CHANGE
Study and Revision Guide

SL AND HL CORE

Simon Oakes

For Sally Eden

Every effort has been made to trace all copyright holders, but if any have been inadvertently overlooked, the Publishers will be pleased to make the necessary arrangements at the first opportunity.

Although every effort has been made to ensure that website addresses are correct at time of going to press, Hodder Education cannot be held responsible for the content of any website mentioned in this book. It is sometimes possible to find a relocated web page by typing in the address of the home page for a website in the URL window of your browser.

Hachette UK's policy is to use papers that are natural, renewable and recyclable products and made from wood grown in sustainable forests. The logging and manufacturing processes are expected to conform to the environmental regulations of the country of origin.

Orders: please contact Bookpoint Ltd, 130 Park Drive, Milton Park, Abingdon, Oxon OX14 4SE. Telephone: (44) 01235 827720. Fax: (44) 01235 400454. Email education@bookpoint.co.uk Lines are open from 9 a.m. to 5 p.m., Monday to Saturday, with a 24-hour message answering service. You can also order through our website: www.hoddereducation.com

ISBN: 9781510403550

© Simon Oakes 2017

First published in 2017 by

Hodder Education,
An Hachette UK Company
Carmelite House
50 Victoria Embankment
London EC4Y 0DZ

www.hoddereducation.com

Impression number   10 9 8 7 6 5 4 3 2 1

Year   2021 2020 2019 2018 2017

All rights reserved. Apart from any use permitted under UK copyright law, no part of this publication may be reproduced or transmitted in any form or by any means, electronic or mechanical, including photocopying and recording, or held within any information storage and retrieval system, without permission in writing from the publisher or under licence from the Copyright Licensing Agency Limited. Further details of such licences (for reprographic reproduction) may be obtained from the Copyright Licensing Agency Limited, Saffron House, 6–10 Kirby Street, London EC1N 8TS.

Cover photo © OJO Images Ltd/Alamy Stock Photo

Illustrations by Aptara, Inc.

Typeset in India by Aptara, Inc.

Printed in Spain

A catalogue record for this title is available from the British Library.

Contents

How to use this revision guide ... iv
Features to help you succeed ... iv
Getting to know the exam ... iv
Assessment objectives ... v
The examination paper and questions ... vi
Describing and identifying patterns, trends and data ... vi
Understanding and using the PPPPSS concepts ... vii

Unit 1 Changing population ... 1
- 1.1 Population and economic development patterns ... 1
- 1.2 Changing populations and places ... 11
- 1.3 Population challenges and opportunities ... 25

Unit 2 Global climate – vulnerability and resilience ... 37
- 2.1 Causes of global climate change ... 37
- 2.2 Consequences of global climate change ... 48
- 2.3 Responding to climate change and building resilience ... 63

Unit 3 Global resource consumption and security ... 76
- 3.1 Global trends in consumption ... 76
- 3.2 Impacts of changing trends in resource consumption – the Water-Food-Energy nexus ... 91
- 3.3 Resource stewardship possibilities ... 103

Glossary ... 116

Acknowledgements ... 119

How to use this revision guide

Welcome to the *Geography for the IB Diploma Revision and Study Guide*.

This book will help you plan your revision and work through it in a methodological way. The guide follows the Geography syllabus for Paper 2 (SL/HL) topic by topic, with revision and exam practice questions to help you check your understanding.

■ Features to help you succeed

PPPPSS CONCEPTS

These 'think about' boxes pose questions that help you consider and consolidate your understanding and application of the key specialized concepts used in Geography – **place**, **process**, **power**, **possibility**, as well as the two further organising concepts of **scale** and **spatial interaction**. Have a go at every question you come across.

Keyword definitions

The definitions of essential key terms are provided on the page where they appear. These are words that you can be expected to define in exams. A **Glossary** of other essential terms, highlighted throughout the text, is given at the end of the book.

EXAM FOCUS

In the Exam Focus sections at the end of each chapter, example answers to exam-style questions are given and reviewed. Examiner comments and mind maps are used to help you consolidate your revision and practice your exam skills.

■ CHAPTER SUMMARY KEY POINTS

At the end of each chapter, a knowledge checklist helps you review everything you have learned over the previous pages. An Evaluation, Synthesis and Skills (ESK) summary is also included. This helps show how the knowledge you have acquired may be applied in order to analyse information, evaluate issues and tackle big geographic questions.

You can keep track of your revision by ticking off each topic heading in the book. Tick each topic heading when you have:
■ revised and understood a topic
■ read the exam-style questions in the Exam focus sections, completed the activities and reviewed any example answer comments.

Use this book as the cornerstone of your revision. Don't hesitate to write in it and personalize your notes. Use a highlighter to identify areas that need further work. You may find it helpful to add your own notes as you work through each topic. Good luck!

Getting to know the exam

Exam paper	Duration	Format	Topics	Weighting	Total marks
Paper 1 options (SL)	2 hours 15 mins	Structured questions and essays	2 options	35	40
Paper 1 options (HL)	2 hours 15 mins	Structured questions and essays	3 options	35	60
Paper 2 core (SL/HL)	1 hour 15 mins	Structured questions and essay	All	40 (SL) 25 (HL)	50
Paper 3 core (HL only)	1 hour 45 mins	Extended writing and essay	All	20	28

At the end of your Geography course you will sit two papers at SL (Paper 1 and Paper 2) and three papers at HL (Paper 1, Paper 2 and Paper 3). These external exams account for 80% of the final marks at HL and 75% at SL. The other assessed part of the course (20% at HL and 25% at SL) is the Internal Assessment, which is marked by your teacher, but externally moderated by an examiner.

Here is some general advice for the exams:
- Make sure you have learned the command terms (describe, explain, suggest, evaluate etc.); there is a tendency to focus on the content in the question rather than the command term, but if you do not address what the command term is asking of you then you will not be awarded marks. Command terms are covered below.
- If you run out of room on the page, use continuation sheets and indicate clearly that you have done this on the cover sheet.
- The fact that the question continues on another sheet of paper needs to be clearly indicated in the text box provided.
- Plan your answer carefully *before* you begin any extended writing tasks.

Assessment objectives

To successfully complete the course, you have to achieve certain assessment objectives. The following table shows all of the commonly-used command terms in Paper 2, which this book supports, along with an indication of the depth required from your written answers.

Describe	AO1	**Assessment objective 1**	These command terms require students to demonstrate knowledge and understanding, for example: 'Identify two greenhouses gases.' They can be used to set skills-based tasks using 'procedural' knowledge too (see page vi).
Identify	AO1	Knowledge and understanding	
Outline	AO1	Part (a) of questions 1, 2, 3 and 4 in your examination will usually include AO1 questions.	
State	AO1		
Explain	AO2	**Assessment objective 2**	These command terms require students to apply their knowledge and understanding to a well-defined task such as: 'Explain political reasons why a country's birth rate may change over time.'
Suggest	AO2	Demonstrate application and analysis Part (a) of questions 1, 2, 3 and 4 in your examination will usually include AO2 questions.	
Discuss	AO3	**Assessment objective 3**	These command terms require students to rearrange a series of geographic ideas, concepts or case studies into a new whole and to provide evaluation or judgements based on evidence. For example: 'To what extent do you agree that the impacts of population growth are always negative?'
Evaluate	AO3	Demonstrate synthesis and evaluation The final parts of Question 4 (a 6-mark piece of extended writing) and Question 5 (a 10-mark essay) are AO3 tasks.	
To what extent	AO3		

The table below defines command words used most commonly in the Geography Paper 2 examination.

Term	Definition
Describe (AO1)	Break down in order to bring out the essential elements or structure.
Outline (AO1)	Give a brief account or summary.
Explain (AO2)	Give a detailed account, including reasons or causes.
Suggest (AO2)	Propose a solution, hypothesis or other possible answer.
Evaluate (AO3)	Make an appraisal by weighing up the strengths and limitations.
Discuss (AO3)	Offer a considered and balanced review that includes a range of arguments, factors or hypotheses. Opinions or conclusions should be presented clearly and supported by appropriate evidence.
To what extent (AO3)	Consider the merits or otherwise of an argument or concept. Opinions and conclusions should be presented clearly and supported with empirical evidence and sound argument.

It is essential that you are familiar with these terms, so that you are able to recognize the type of response you are expected to provide.

The examination paper and questions

You must answer all of the short-answer questions in Sections A and B of the examination paper. Examples of these questions appear on pages 10, 25, 62 and 91 (Section A questions) and page 114 (Section B questions). You must also choose one from two optional questions in Section C of the examination paper.

The Section C essay task asks to what extent you agree or disagree with a generalized statement. It is an AO3 task, which requires you to write critically and in a conceptually informed way about the question statement. You are required to argue for and against a viewpoint, using carefully chosen evidence and ideas, before arriving at a conclusion. You should spend around 15–20 minutes on this task. Examples of Section C questions include:

(b) 'The benefits of demographic transition over time are greater than any costs.' To what extent do you agree with this statement? [10 marks]

(b) 'The world's poorer countries are least responsible for climate change.' To what extent do you agree with this statement? [10 marks]

A simplified version of the levels-based mark scheme for these questions looks like this:

■ Mark scheme

1–4 marks	Response is general, not focused on the question, and lacks detail and structure. No synthesis or evaluation is expected.
5–6 marks	Response only partially addresses the question, with limited subject knowledge links. Evidence is both relevant and irrelevant, and is largely unstructured. No synthesis or evaluation is expected.
7–8 marks	Response addresses most parts of the question, with developed subject knowledge links. Ideas are supported by relevant evidence. A structured synthesis of different ideas *or* a genuine evaluation is required.
9–10 marks	Balanced response addresses all aspects of the question, ideas are well explained and there is sustained use of well-integrated evidence. Both a well-structured synthesis *and* a genuine evaluation are required.

Describing and identifying patterns, trends and data

A significant number of marks in your examination – typically between 6 and 8 – are reserved for tasks that require you to use 'procedural' knowledge. This means knowing how to carry out an AO1 task such as:
- identifying a maximum or minimum value in a data set
- describing a pattern or distribution on a global or more localized map
- describing one or more trends on a graph showing changes over time.

These tasks do not require you to explain the data you have been shown. However, you do need to make use of your own 'procedural' knowledge. Geographers generally follow certain procedures when describing data, as the table below shows.

Procedures to follow when describing a pattern on a map	Procedures to follow when describing a trend on a graph
Offer an overview: is the general picture one of an uneven or even distribution?	Offer an overview: is the general picture one of a rise or fall over time?
Refer to points of the compass, or lines of latitude, if this brings clarity.	Refer to the rate of any changes: how steep are the lines on the graph?
If there is a scale, make use of it by including estimated distances in your answer, if this brings clarity.	Refer to the evenness or unevenness of any trend: are there any fluctuations, or are the lines smooth?
Refer to named places (cities, countries or continents).	Use data in your answer: identify maximum or minimum values if this brings clarity.
Refer to data (if values have been plotted at points on the map).	Draw attention to any anomalies (values that deviate from the general trend in some way).
Draw attention to any anomalies (places or values that stand out as being different in some way).	Distinguish between any actual (past) and projected (future) trends.

Understanding and using the PPPPSS concepts

An important feature of the new 2017 Geography course is the inclusion of specialized and organising Geographic concepts. These are shown in the diagram below.

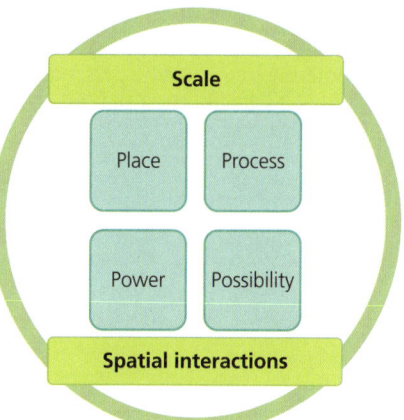

Place	A portion of geographic space that is unique in some way. Places can be compared according to their demographic characteristics, or disparate exposure to climate change risks. The characteristics of a place may be real or perceived.
Process	Human or physical mechanisms of change, such as migration or sea-level changes. Processes operate on varying timescales. Linear systems, circular systems and complex systems are all outcomes of the way in which processes operate and interact.
Power	The ability to influence and affect change or equilibrium at different scales. Power is vested in citizens, governments, institutions and other players, and in processes in the natural world. Equity and security, both environmental and economic, can be gained or lost as a result of the interaction of powerful forces.
Possibility	Alternative events, futures and outcomes that geographers can model, project or predict with varying degrees of certainty. Key contemporary possibilities include the degree to which human and environmental systems are sustainable and resilient, and can adapt or change.
Scale	Places can be identified at a variety of geographic scales, from local territories to the national or state level. Climate change affects the world at a planetary level.
Spatial interactions	Flows, movements or exchanges that link places together. Interactions such as migration may lead to two places becoming interdependent on one another.

Essentially, these six ideas help provide you with a roadmap to 'thinking like a geographer', when tackling an essay question, such as the following example:

To what extent do you agree that the impacts of population growth are always negative?

This is very broad question and as a result potentially tricky to answer well. The specialized and organising Geographic concepts can be used to help you 'scaffold' your answer. Familiarity with the PPPPSS framework provides you with the basis for a series of further questions you may want to address as part of your overall answer, as follows:

- Have many **places** and societies ever suffered negative impacts of population growth, such as famine? *(This may prompt you to write about Malthusian theory and case studies.)*
- What **processes** lead to population growth? *(This may prompt you to write about population growth resulting from both migration and natural increase in your answer.)*
- Do some countries and governments have the **power** to cope with population growth? *(This may prompt you to write about examples of population policies.)*
- Is there a **possibility** that population growth can have positive impacts for places? *(This may prompt you to write about the demographic dividend, or Water-Food-Energy nexus issues.)*
- Are some local-**scale** communities particularly affected by population growth? *(This may prompt you to write about the impacts of population growth on cities and megacities.)*
- What different kinds of **spatial interaction** can result from population growth? *(This may prompt you to write about positive and negative impacts of out-migration from countries and regions where population is growing rapidly, such as Europe during its demographic transition.)*

In order to write a good essay, you do not need to do all of this, of course. But it may be helpful to try and draw on two or three of the specialized and organizing concepts when planning an essay. You can also make use of the important Group 3 idea of **perspectives**. Most questions can be debated from the varying and sometimes contrasting perspectives of different **stakeholders** or players (all of whom may be located at either local, national or global scales).

Unit 1 Changing population

1.1 Population and economic development patterns

Revised

The study of population patterns and dynamics is a foundation topic for Geography. The purpose of this first chapter is to provide you with a broad, factual and conceptual introduction to population studies. This chapter also introduces the key concept of **development** and explores how levels of development vary spatially at different geographical **scales**.

Population distribution at the global scale

Revised

Population distribution and **population density** patterns can be investigated at varying spatial scales. Figure 1.1 shows the global distribution of population. Important features include the following:

- People are distributed unevenly among the world's continents. Over half of Earth's population is in Asia; 1.2 billion people live in Africa and a further billion are spread across North and South America. The figures for Europe and Oceania (Australia and New Zealand) are 740 million and 40 million respectively (2016 data).
- Many people live along coastlines; relatively fewer inhabit the continental interiors. This is one reason why the **Lorenz curve** for global population distribution looks the way it does (Figure 1.2).
- Just one-third of Earth's surface is land and more than two-thirds of this is inhabited by fewer than 20 persons per square kilometre, including (when looking at the national average) Russia, Canada, Australia, Greenland, most of South America, Antarctica and Saharan Africa.

> **Keyword definitions**
>
> **Development** Human development generally means the ways in which a country seeks to progress economically and also to improve the quality of life for its inhabitants. A country's level of development is shown firstly by economic indicators of average national wealth and/or income, but can encompass social and political criteria, too.
>
> **Scale** Places, areas or territories can be studied and identified at a variety of geographic scales, from local territories to the national or state level. The global distribution of population is a macroscale (planetary scale) data pattern. In contrast, very small-scale patterns are sometimes called microscale distributions.

Figure 1.1 Dot map showing the global distribution of areas where there is a high density of population

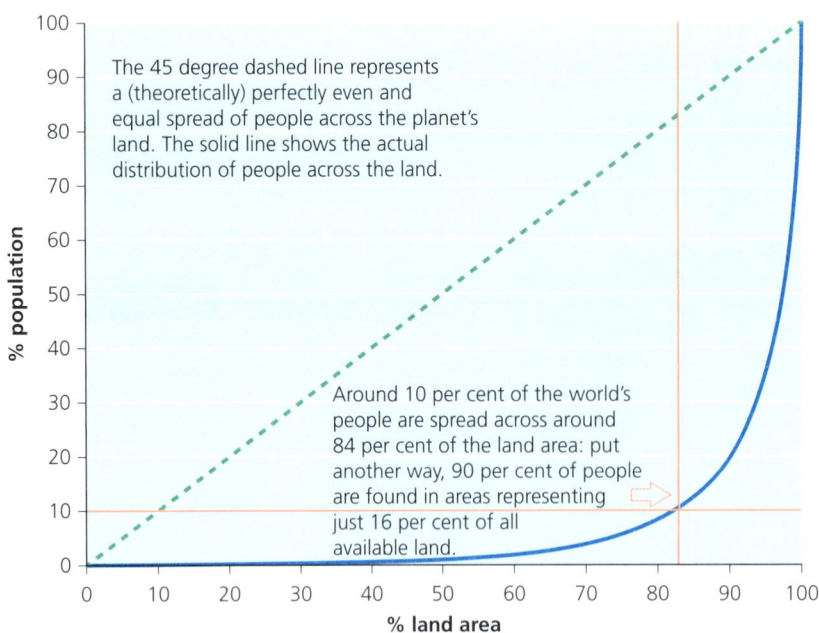

Figure 1.2 A Lorenz curve showing the unequal distribution of the world's population (Table 1.1)

Physical and human factors affecting global population distribution

Both physical and human factors affect population distribution patterns at the global scale (Table 1.1). Historically, early settlers in any world region lived, either by choice or necessity, wherever the environment provided them with 'a foothold to livelihood'. Even today, around one-third of the world's economically active population obtains its food and/or income by actively farming the land. This means that physical influences on food production – including climate and soil fertility – remain hugely important factors in determining where more than 2 billion of the world's people live and work.

- This partly explains the low levels of density in continental interiors: inaccessibility and extremes of climate (including high daily or annual temperature ranges) mostly discourage large-scale settlement in central areas of Asia and Saharan Africa.
- Historically, fewer people have settled permanently in regions where water availability is lacking for all or part of the year, such as the Sahara and Gobi deserts. In contrast, a linear distribution of population can be observed in Figure 1.1 following the course of major rivers including the Nile and Amazon.
- Population tends to be sparse in mountainous regions such as the Tibetan plateau and American Rockies.

In the absence of technology, physical factors help establish whether or not a region will become home to a significant human population. Over time, however, imbalances between regions may become amplified or lessened on account of human factors.

- For instance, hot and dry climates may attract large numbers of settlers once sufficient capital and technology are available to provide water supplies artificially. Between 2000 and 2010, several states in the USA's **arid** southwest experienced rapid population growth rates more than double the national average. These were Nevada (35 per cent), Arizona (25 per cent) and Utah (24 per cent). This rapid growth has been sustained by pipeline transfers of water from the Colorado River.
- The uneven distribution of mineral deposits and fossil fuels can help explain pockets of prosperity in areas where population is generally sparse as a result of climatic factors. The growth of large urban areas in the Middle East – including Riyadh in Saudi Arabia and Doha in Qatar – has been made possible by the oil wealth, which pays for air conditioning and the **desalinization** of seawater.

PPPPSS CONCEPTS

Use the concept of scale to analyse population patterns in your own country. Begin by thinking about how people are distributed within your local neighbourhood; then analyse the pattern at a larger scale, for example in your country as a whole.

Keyword definitions

Population distribution A description of the way in which people are spread out across the Earth's surface. For instance, around 4 billion people live in Asia.

Population density The number of people living within a specified area. For instance, the population density of large parts of New Mexico (USA) is less than one person per square kilometre.

Lorenz curve A diagrammatic expression of the extent to which a distribution is unequal. The dashed straight diagonal line on a Lorenz curve shows a perfectly even and equal distribution. The further away the solid curved line deviates from this dashed line, the greater the level of inequality that actually exists for the scenario shown.

Arid A climate whose precipitation is less than 250mm annually.

Desalinization The removal of salt water and other minerals from seawater. The process is costly and requires desalinization plants to be built.

Table 1.1 How physical and human factors can influence large-scale variations in population distribution

		Sparse population	**Dense population**
Physical factors	Physical accessibility	Rugged mountains (Alps) High plateaux (Tibet)	Flat lowlands (Netherlands, Nile valley)
	Relief and soils	Frozen soils (Siberia) Eroded soils (Sahel)	Deep humus (Paris basin) River silt (Ganges delta)
	Climate	Low temperatures (Canada and Alaska)	Longer growing season (tropical Asia)
	Vegetation	Dense forest which restricts human activity (Amazonia)	Grassland ecosystems (eastern Europe)
	Water supply	Insufficient or unpredictable supply (Australian desert)	Mostly reliable all-year supply (Western Europe)
Human factors	Economic factors	Extensive agriculture (few workers needed per unit area)	Ports (Singapore) Intensive farming (China)
	Political factors	Low levels of state investment (interior of Brazil)	Forced movements (Soviet settlement of Siberia)
	Technological factors	Lack of technology needed to increase water availability	Irrigation and desalinization technologies available

■ Hazards, resources and human settlement

Another useful way of investigating population distribution is to think critically about the following hypothesis: 'People are most likely to live where resources are maximized and hazards are minimized.' Is this statement broadly true? In fact, the relationship is far more complex than it may appear initially.

- Firstly, it is important to recognize that hazards and resources are, essentially, two sides of the same coin. Both are best described as 'relationships' between humans and the natural world. The only difference is that resources are a 'positive' interaction and hazards are a 'negative' interaction (Figure 1.3).
- Secondly, many things found in the natural environment are both a hazard *and* a resource: rivers and coastlines function as sources of water, food or transport and yet can be incredibly dangerous to live next to. Many of the world's highest-density pockets of population are located in tectonically hazardous areas. This is because they have been drawn to coastlines that have formed along continental plate boundaries. The Californian coastline is a good example of this: the economic success of Los Angeles and San Francisco owes much to their position on the Pacific coastline. However, these are also dangerous places to live on account of the high earthquake risk.
- Relationships between people and the environment are constantly changing because of technology. We can build defences to protect ourselves from river flooding while continuing to make use of its water as a resource. Thanks to water transfer schemes, the city of Las Vegas currently prospers in a desert area that was once viewed as a life-threatening environment.

In summary, the relationship between physical factors and population distribution is complex, dynamic and mediated by technology. It is best to think carefully before making sweeping and deterministic generalizations that attempt to link the presence or absence of people with particular types of hazard or climate.

> **PPPPSS CONCEPTS**
>
> Think about how human and physical factors interact in ways which give rise to uneven population patterns at continental and global scales.

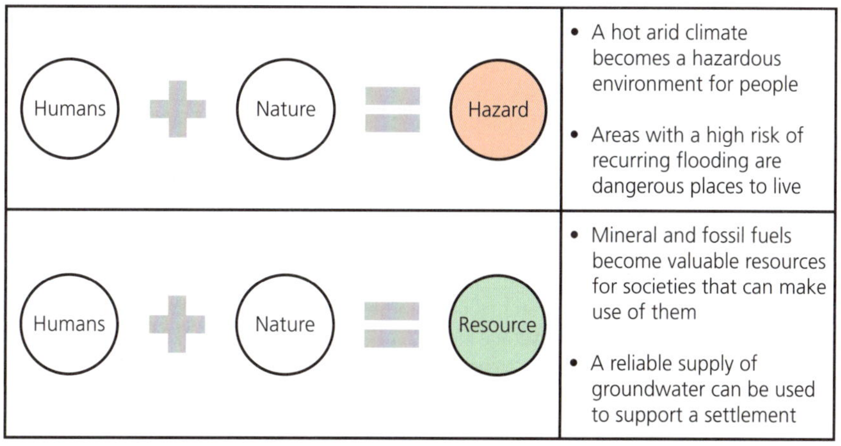

Figure 1.3 Hazards are harmful interactions between humans and the natural world; resources are beneficial interactions

- Humans + Nature = Hazard
  - A hot arid climate becomes a hazardous environment for people
  - Areas with a high risk of recurring flooding are dangerous places to live
- Humans + Nature = Resource
  - Mineral and fossil fuels become valuable resources for societies that can make use of them
  - A reliable supply of groundwater can be used to support a settlement

Global patterns and classification of economic development

Revised

Becoming familiar with the world development map is another essential foundation step for studying Geography. The uneven pattern of global development is shown in Figure 1.4. This shows the world divided crudely into a **core** of high-income developed countries, a **semi-periphery** comprised of middle-income countries (or emerging economies) and a **periphery** of low-income developing countries. The characteristics of these three global groups are outlined further in Table 1.2.

The distribution pattern for emerging economies (EEs) and low-income countries (LICs) is complicated and is changing constantly. Key features are that:

- most South American countries are EEs
- Asia now has more EEs than LICs
- Africa still has more LICs than EEs
- eastern European countries (including some European Union members, are classified mainly as EEs).

The global pattern of economic development has changed radically over time. In the 1980s, there was still a clear divide between the rich 'global north' and the poor 'global south'. This crude division now appears hopelessly outdated, especially when you consider that:

- China has become the world's largest economy
- several of the world's highest-income countries, including Qatar, Kuwait and Singapore, are part of what used to be called 'the global south'
- the average income of some European populations, including Hungary and Bulgaria, is lower than that of either Brazil or Malaysia
- large numbers of millionaires and billionaires can be found in every populated continent, including Africa.

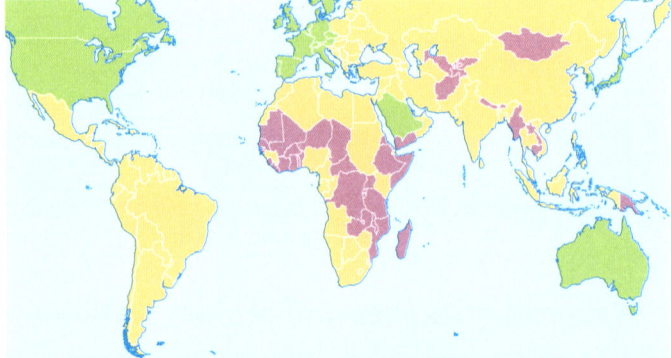

Key		World Bank category	Average income (US$)
	Low-income countries	Low income	1045 or below
	Emerging economies	Middle income	1046–12,735
	High-income countries	High income	12,736 or more

Figure 1.4 The world map of development (World Bank data)

Table 1.2 The three main groups of countries (classified by per capita income and economic structure)

Low-income developing countries *Global periphery*	This group of around 30 countries is classified by the World Bank as having low average incomes of US$1,045 or below (2015 values) and a low per capita gross domestic product (GDP). Agriculture still plays an important role in their economies. Some of these states suffer from political instability and conflict; some, including Somalia and Eritrea, have been described pejoratively as 'failed states' (the term 'fragile state' may be preferable to use).
Middle-income emerging economies (EEs) *Semi-periphery*	These are around 80 countries that have begun to experience higher rates of economic growth, usually due to rapid factory expansion and industrialization. The number of EEs has increased rapidly in recent decades: this is linked to the spread of **globalization** and investment by **transnational corporations (TNCs)** in EEs (leading to economic 'take-off' being achieved). Emerging economies are roughly synonymous with the World Bank's 'middle-income' category of countries and include China, India, Indonesia, Brazil, Mexico, Nigeria and South Africa. They are home to a rising number of the 'global middle class' (people with discretionary income they can spend on consumer goods). Definitions of this vary: some organizations define the global middle class as people with an annual income of over US$10,000; others use a benchmark of US$10 per day income.
High-income developed countries (HICs) *Global core*	This group of around 80 countries is classified by the World Bank as having high average incomes of US$12,736 or above (2015 values). Around half are sometimes called 'developed' countries. These are states where office work has overtaken factory employment, creating a 'post-industrial' economy. There are also around 40 smaller high-income countries and territories (of roughly one million people or fewer), including Bahrain, Qatar, Liechtenstein and the Cayman Islands.

Keyword definitions

Globalization The variety of accelerating ways in which places and people have become connected with one another as part of a complicated global system.

Transnational corporations (TNCs) Businesses whose operations are spread across the world, operating in many nations as both makers and sellers of goods and services. Many of the largest are instantly recognizable 'global brands' that bring cultural change to the places where products are consumed.

Measuring economic development using gross domestic product (GDP)

Gross domestic product (GDP) is a widely used measurement of national economic wealth that can be used to map spatial variations in economic development. It is one of the best-known measures of national and global prosperity used by the World Bank, which recently estimated global GDP in 2014 at about US$78 trillion.

Calculating GDP is not an easy task. Numerous earnings of citizens and businesses need to be accounted for, not all of which are easily recordable. Work in the **informal sector** of employment is notoriously hard to quantify. Table 1.3 shows the numerous steps that are taken to calculate GDP data.

Table 1.3 Quantifying a country's economic development in four steps using the GDP formula

Step	
1	Individual countries make a calculation of their GDP each year, using globally agreed guidelines (overseen by the United Nations). Data are generated using a country-specific formula which establishes the weighting given to different economic sectors, such as agriculture and industry. This same weighting – or set of sums – is re-applied every year until it is felt the formula should be changed (for instance, when a country is developing rapidly).
2	The figures are then verified by international organizations such as the International Monetary Fund, the World Bank or the African Development Bank.
3	The next step is to convert all of the national GDP estimates into a common currency, US dollars, in order for comparisons and ranking (however, the volatility of exchange rates means that some GDP data may quickly become an unreliable way of comparing countries).
4	Economists believe that, at the same time as they are converted into US dollars, each country's GDP data should additionally be manipulated so that an estimate of the real cost of living, known as **purchasing power parity** or PPP, is factored in. Simply put, in a low-cost economy, where goods and services are relatively affordable, the GDP should be increased and vice versa. This is why, if you consult Wikipedia, you will find two estimates given of every country's GDP. Brazil, for instance, has a 'nominal' GDP of US$2.14 trillion in 2017, and a 'PPP' GDP of US$3.22 trillion. This suggests that the price of, say, a Big Mac, is relatively cheaper in Rio than it is in New York – can you see why this is the case?

While there is some obvious merit in using GDP to measure development, there are considerable grounds for criticism too. During 2014, Nigeria's government doubled its GDP 'overnight' to become Africa's largest, ahead of South Africa. Had Nigeria suddenly become richer?

- In fact, what had happened was a 're-basing' of the GDP formula.
- Up until then, Nigeria's national film industry, known as Nollywood, had been excluded from the country's GDP data. This was because the GDP formula has been written in 1990 when the structure of Nigeria's economy was very different and the film industry, banking and telecoms had barely taken off.
- The Nigerian government therefore decided to reduce agriculture's share of GDP from 35 per cent to 22 per cent; contrastingly, the weight of telecoms was increased from just 1 per cent of GDP to 9 per cent. Nollywood earnings have been included in the calculation for the first time.
- As a result, Nigerian GDP for 2013 was US$509 billion, 89 per cent larger than the previous year.

> **Keyword definitions**
>
> **Gross domestic product (GDP)** A measure of the total value of the output of final goods and services inside a nation's borders. Each country's annual calculation includes the value added by any foreign-owned businesses that have located operations there.
>
> **Informal sector** Unofficial forms of employment that are not easily made subject to government regulation or taxation. Sometimes called 'the black economy' or 'cash in hand' work, informal employment may be the only kind of work that poorly educated people can get.
>
> **Purchasing power parity** A measure of average wealth that takes into account the cost of a typical 'basket of goods' in a country. In low-income countries, goods often cost less, meaning that wages go further than might be expected in a high-income country.

> **PPPPSS CONCEPTS**
>
> Think about the processes by which data are created. How reliable and valid are the data we use to study people and economies? Do the data perhaps tell us more about the process that created them than they do about any 'real' or measurable phenomenon?

Variations in population and development at the national scale

Revised

Within all countries, population density varies between local regions, resulting in an uneven population distribution. In some cases, the differences are extreme: 99 per cent of Egypt's people live in the Nile valley (just 4 per cent of the total land area). More than 30 per cent of the UK's population are in the southeast of

England living at a density level of more than 300 persons square kilometre. In contrast, just 4 million people inhabit Scotland, despite its much larger area. This population imbalance is both caused by, and further creates, a national economic balance.

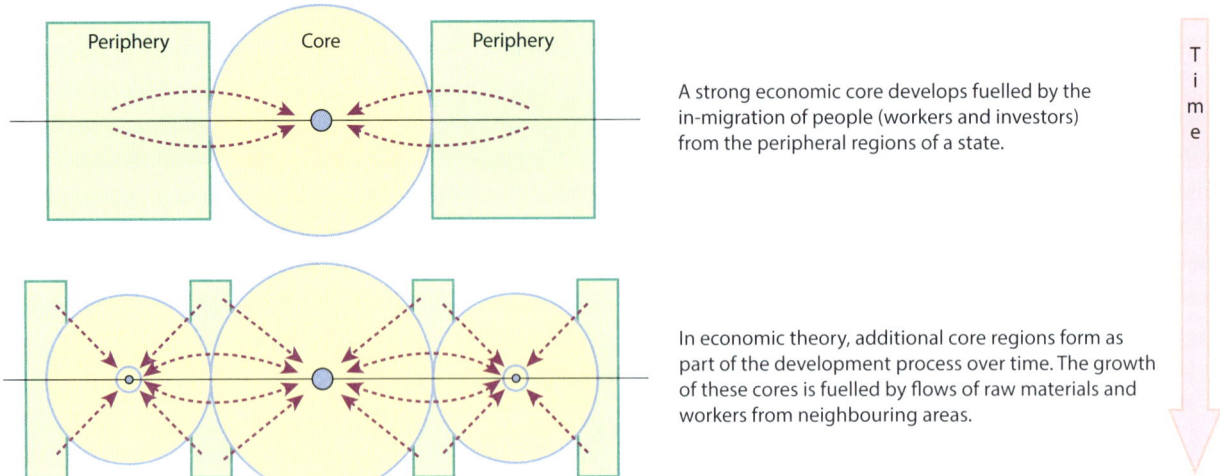

Figure 1.5 Backwash processes in the Friedmann core-periphery model

Regional imbalances arise initially within countries because of physical factors and the natural advantages of certain locations over others (such as raw material availability or the presence of a coastline). Subsequently, these favoured areas become 'national-scale core' regions. They build on their natural advantage through time: the centripetal (attracting) force they exert draws in ambitious migrants, investors and resources from other regions. This virtuous circle of spiralling growth was called 'cumulative causation' by the theorist Gunnar Myrdal. The inflows of migrants and resources to the core are called **backwash** effects (Figure 1.5).

The result of cumulative causation is the development of national **core-periphery systems** that keep strengthening over time on account of positive feedback effects.

Sometimes, this process of core-periphery polarization can be seen operating at larger spatial scales than the state level. Within the European Union (EU), free movement of labour has helped an international core-periphery pattern to develop. The EU core region encompasses southern England, northern France, Belgium and much of western Germany. It includes the world cities of London, Paris, Brussels and Frankfurt. Labour migration flows from eastern and southern Europe are directed overwhelmingly towards these places.

Voluntary internal migration

Much of the voluntary **internal migration** that takes place within core-periphery systems is directed from rural to urban areas. This rural–urban migration is the result of push and pull factors and is encouraged further by the availability of good transport links and communications.

- In terms of the numbers involved, rural–urban migration is the most significant population movement occurring globally. Within a few years, there will be one billion rural–urban migrants living in the world's towns and cities.
- Global **urbanization** passed the threshold of 50 per cent in 2008, meaning that the majority of people now live in urban areas. By 2040, it is expected that over 70 per cent of the world's population will live in towns and cities compared with less than 30 per cent in 1950. Table 1.4 and Figure 1.6 explain why rapid urbanization is still happening in many places.

Keyword definitions

Backwash Flows of people, investment and resources directed from peripheral to core regions. This process is responsible for the polarization of regional prosperity between regions within the same country.

Core-periphery system The uneven spatial distribution of national population and wealth between two or more regions of a state or country, resulting from flows of migrants, trade and investment.

Internal migration The movement of people from place to place inside the borders of a country. Globally, most internal migrants move from rural to urban areas ('rural–urban' migrants). In the developed world, however, some people move from urban to rural areas too (a process called counterurbanization).

Urbanization An increase in the proportion of people living in urban areas.

1.1 Population and economic development patterns

Table 1.4 Causes of rural–urban migration

Urban pull factors	The main factor almost everywhere is employment. Foreign direct investment (FDI) by TNCs in cities of emerging economies provides a range of work opportunities with the companies and their supply chains. We can distinguish between formal sector employment (working as a salaried employee of Starbucks in São Paulo, for instance) and the informal sector (people scavenging material for recycling at landfill sites in Lagos). Urban areas offer the hope of promotion and advancement into professional roles that are non-existent in rural areas. Additionally, schooling and healthcare may be better in urban areas, making cities a good place for young migrants with aspirations for their children.
Rural push factors	The main factor is usually poverty, aggravated by population growth (not enough jobs for those who need them) and land reforms (unable to prove they own their land, subsistence farmers must often relocate to make room for TNCs and cash crops). Agricultural modernization reduces the need for rural labour further (including the introduction of farm machinery by global agribusinesses like Cargill). Resource scarcity in rural areas with population growth, such as the Darfur region of Sudan, may trigger conflict and migration (people then become classified as refugees and not economic migrants, however).
'Shrinking world' technology	Rural dwellers are gaining knowledge of the outside world and its opportunities. The 'shrinking world' technologies we associate with globalization all play important roles in fostering rural–urban migration. Satellites, television and radio 'switch on' people in remote and impoverished rural areas. As people in Africa and Asia begin to use inexpensive mobile devices, knowledge is being shared. Successful migrants communicate useful information and advice to new potential migrants. Also, transport improvements, such as South America's famous Trans-Amazon Highway, have removed **intervening obstacles** to migration.

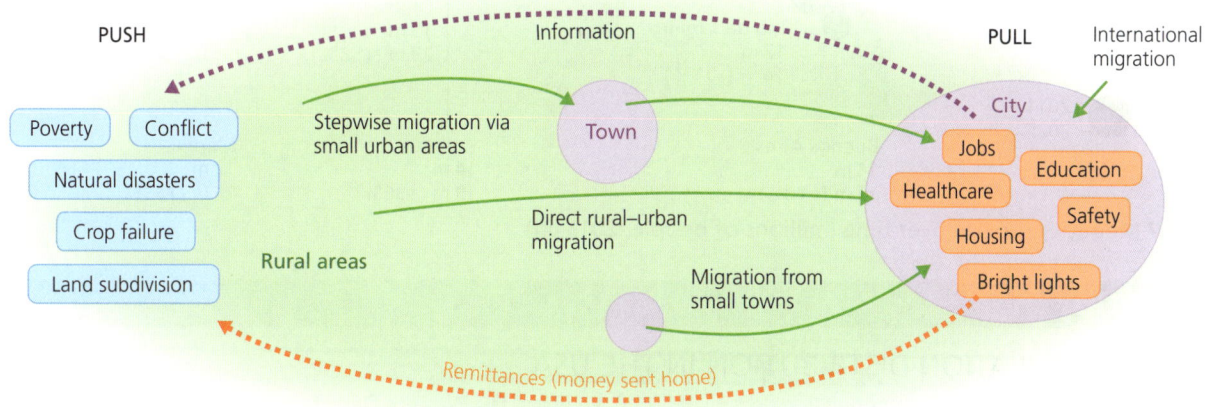

Figure 1.6 A model of rural–urban migration

Megacity growth and national disparities

A megacity is home to 10 million people or more. In 1970 there were just three; by 2020 there will be 30. They grow through a combination of rural–urban migration and natural population increase due to large numbers of children being born (on account of the fact that many migrants are young adults of child-rearing age).

Megacity growth can give rise to marked disparities in terms of how a nation's population is distributed:
- Almost one third of Japan's 128 million people live in Tokyo and its surrounding metropolitan area.
- One-sixth of Mexico's 120 million people live in Mexico City and its surrounding metropolitan area.

Megacities in low-income (developing) and middle-income (emerging) countries have grown especially rapidly (Figure 1.7). São Paulo gains half a million new residents annually from migration. New growth takes place at the fringes of the city where informal (shanty) housing is built by the incomers. Centripetal migration brings people to municipal dumps (Lagos), floodplains (São Paulo), cemeteries (Cairo) and steep, dangerous hill slopes (Rio). Over time, informal housing areas may consolidate as expensive and desirable districts. Rio's now-electrified shanty town Rocinha boasts a McDonald's, hair salons and health clinics.

International migration continues to bring population growth, albeit far more slowly, to megacities in the developed world (for example, eastern Europeans moving to greater London, or Mexicans to Los Angeles). There is residual internal

> **Keyword definition**
> **Intervening obstacles** Barriers to a migrant such as a political border or physical feature (deserts, mountains and rivers).

migration taking place in developed countries too, for instance from the rural heartlands of the USA to New York.

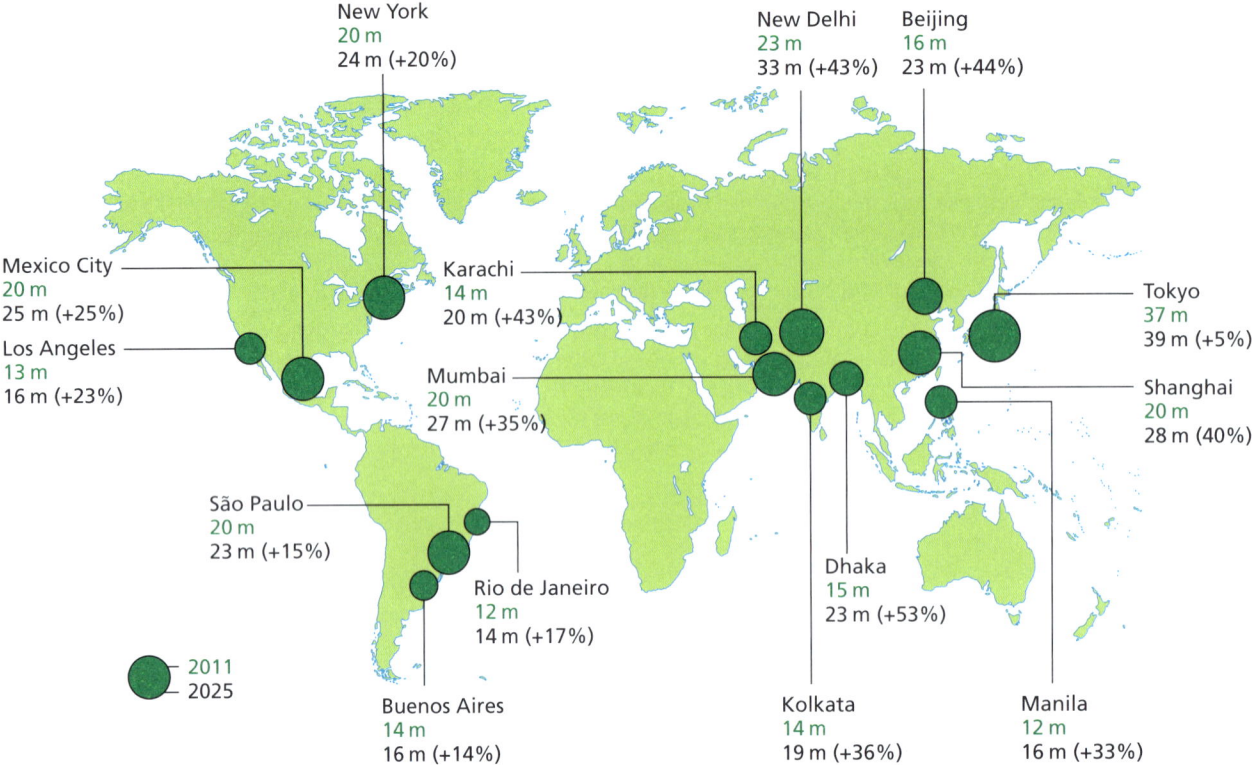

Figure 1.7 Megacity growth over time (millions of people, 2011–15)

CASE STUDY

UNEVEN POPULATION DISTRIBUTION IN THE USA

Overall, the USA has a low average population density of 33 persons per square kilometre. The figure varies greatly from place to place within the USA, however. High densities are found along the eastern coast, especially in the cities of New York, Philadelphia and Baltimore. This is, in part, a legacy of the way Europeans founded new settlements along the eastern seaboard in the 1600s and 1700s. Subsequently, economic activity has remained focused on the eastern coastline and the opportunities for trade it brings.

Over time, early European settlers moved westwards, often displacing the **indigenous populations** of each territory they entered. Large clusters of population developed historically around the Great Lakes in cities such as Detroit and Chicago. However, much of the interior of the USA has a low density of fewer than 20 people per square kilometre. In many areas west of the Rocky Mountains the figure is less than one person per square kilometre. This reflects the aridity of the Western Desert, including large parts of Texas, New Mexico, Nevada and Arizona. In recent decades, some cities in this region, such as Phoenix, have gained population thanks to irrigation and water transfer schemes. Much of the area remains devoid of settlement and population though.

Finally, the west coast – from San Diego as far north as Seattle – is more densely populated, especially the conurbations surrounding the major cities of Los Angeles and San Francisco. Many migrants were attracted here on account of its hot, dry climate and the opportunities provided by the Pacific coastline for trade with Asia. Los Angeles has become a global entertainment industry hub, which is home to Hollywood. The USA is sometimes said to display a 'binary' settlement pattern: the world cities of New York and Los Angeles are equally powerful magnets for migrants and investors alike.

> **Keyword definition**
>
> **Indigenous population** An ethnic group that has occupied the place where it lives and calls home for hundreds or thousands of years without interruption.

Figure 1.8 The uneven population distribution of the USA

CASE STUDY

UNEVEN POPULATION DISTRIBUTION IN CHINA

China is home to around 1.3 billion people. This is four times greater than the population of the USA. Yet, despite this contrast in the number of people, some similarities in their distributions, are obvious. The highest concentrations of people are found along China's east coast. Density reaches a maximum in the province of Jiangsu, which is home to many of the world's leading exporters of electronic equipment and has been China's largest recipient of foreign investment over many years. Further south, an urban mega-region of 120 million people has grown around the Pearl River Delta. It includes the conjoined cities of Shenzhen, Dongguan and Guangzhou. Hong Kong and Macao also form part of this region. Although both territories have been returned to China, they were formerly under British and Portuguese control respectively and have long been important trade hubs where many people have wanted to live.

Population density falls markedly towards the west in China. This reflects both human factors (the reduced potential for international trade with distance from the coastline) and physical factors. Parts of China's interior are extreme environments. The Tibetan plateau is a high-altitude region covering 2.5 million square kilometers, where temperatures fall as low as –40°C in the winter months. The Gobi desert is a vast, sparsely populated area that overlaps part of northern China.

Political influences

In the last 40 years, major population relocations have taken place, which have reduced the population density of some provinces while increasing the density of others markedly. China's total population has not grown very rapidly on account of (until recently) strict political controls limiting the number of births (see page 30). Therefore, population changes in most regions are largely attributable to migration.

- Since 1978, when political and economic reforms began in China, more than 300 million people have left rural areas in search of a better life in cities.

- Only a strict registration system called *hukou* has prevented rural villages from emptying altogether. During this time, the percentage of China's population living in cities has risen from 20 to nearly 60 per cent.

- Between 1990 and 2010, the population of Guangdong province in the Pearl River Delta grew from 62 million to 104 million: a phenomenal rise over a 20-year period.

China's population distribution is shown in Figure 1.9 using a choropleth map. This is a type of map which uses differences in shading or colouring to indicate the average value (in this case of population density) found in a particular area or territory. Figure 1.8 is also a choropleth map.

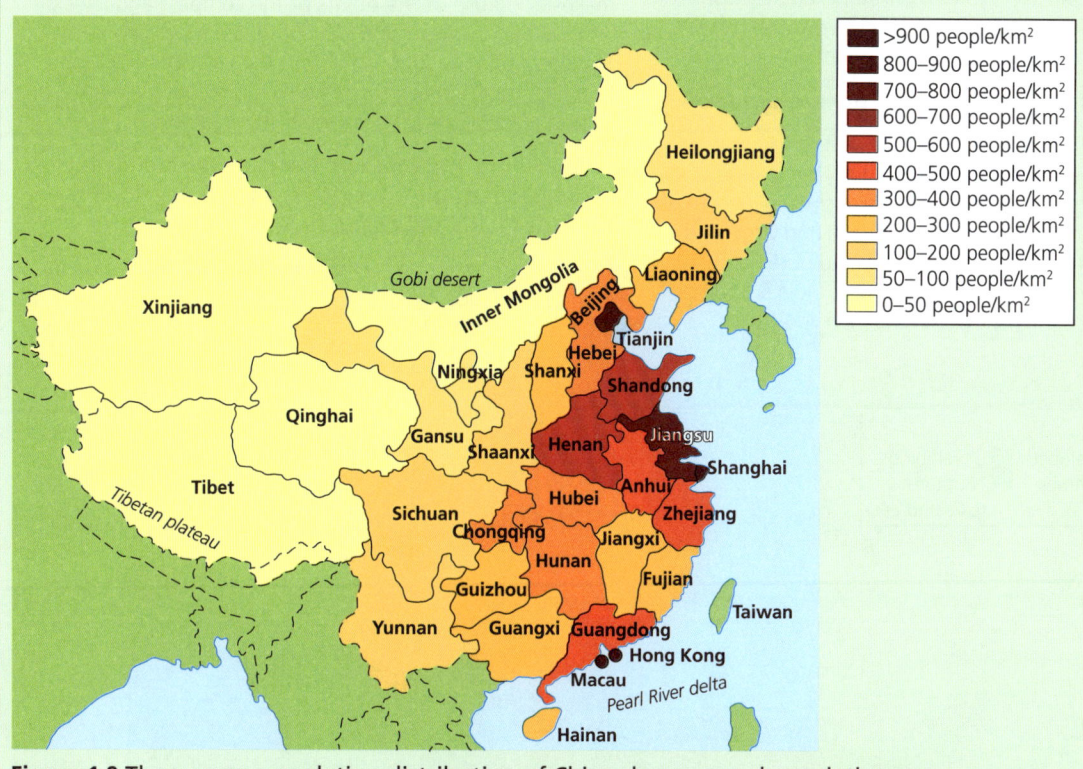

Figure 1.9 The uneven population distribution of China shown as a choropleth map

PPPPSS CONCEPTS

Think about the strengths and weaknesses of Figures 1.8 and 1.9 as ways of showing how population is distributed between different places within a country. Can you suggest any improvements?

Unit 1 Changing population

■ KNOWLEDGE CHECKLIST:

- The global pattern of population distribution
- The influence of physical and human factors on global population distribution, including hazards and resources
- The concept of economic development
- The global pattern of economic development
- The classification of economic development using GDP per capita
- Core–periphery patterns of population and economic development at the national scale
- The process of voluntary internal migration (rural–urban migration)
- Megacity growth and its impacts on national disparities
- Two contrasting examples of uneven population distribution (USA, China)

EVALUATION, SYNTHESIS AND SKILLS (ESK) SUMMARY:

- How physical and human influences on population patterns vary in their relative importance
- How population and development patterns can be studied at varying scales
- How interactions between rural and urban places help explain uneven patterns of population and development

EXAM FOCUS

DESCRIBING PATTERNS AND TRENDS

Once you have acquired knowledge and understanding of a topic, your course may require you to apply what you have learned to stimulus material, such as a map or chart. Each structured question in Section A of your examination begins in this way.

It is essential that you have mastered the skills needed to describe patterns and trends competently. This is because structured questions usually begin with a skills-based task.

Below are sample answers to two part (a) exam-style short-answer questions that use the command word 'describe'. Read them and the accompanying comments.

Study Figure 1.1 (page 1). Describe the pattern shown. [3 marks]

The pattern is extremely uneven overall. Very large areas of the Earth's land surfaces are underpopulated, including large parts of Central Asia and Northern Russia, Canada and Greenland. The highest densities of population are found north of the Tropic of Cancer in Eastern Asia and Europe. India is another highly populated country shown on the map. In Africa, South America and North America population is distributed along the coastlines, with far fewer people in the continental interiors. One exception to this is the line of population which follows the Nile Valley from north into Central Africa.

Examiner's comment

This is a well-articulated answer which includes the *most important features* of the distribution. The description does more than simply naming or listing countries. It also makes good use of the points of the compass and lines of latitude to convey an *overview* of where most people live. The description is supported with the use of *terminology* (such as 'continental interior') which provides clarity and an 'expert voice'. Overall, this would score full marks.

Study Figure 1.10 (below). Describe the trends for (1) developed countries (HICs) and (2) developing countries (LICs and EEs). [2 + 2 marks]

The most important trend shown is an overall increase in world population from just under 3 billion to nearly 8 billion between 1945 and 2020. Growth was most rapid

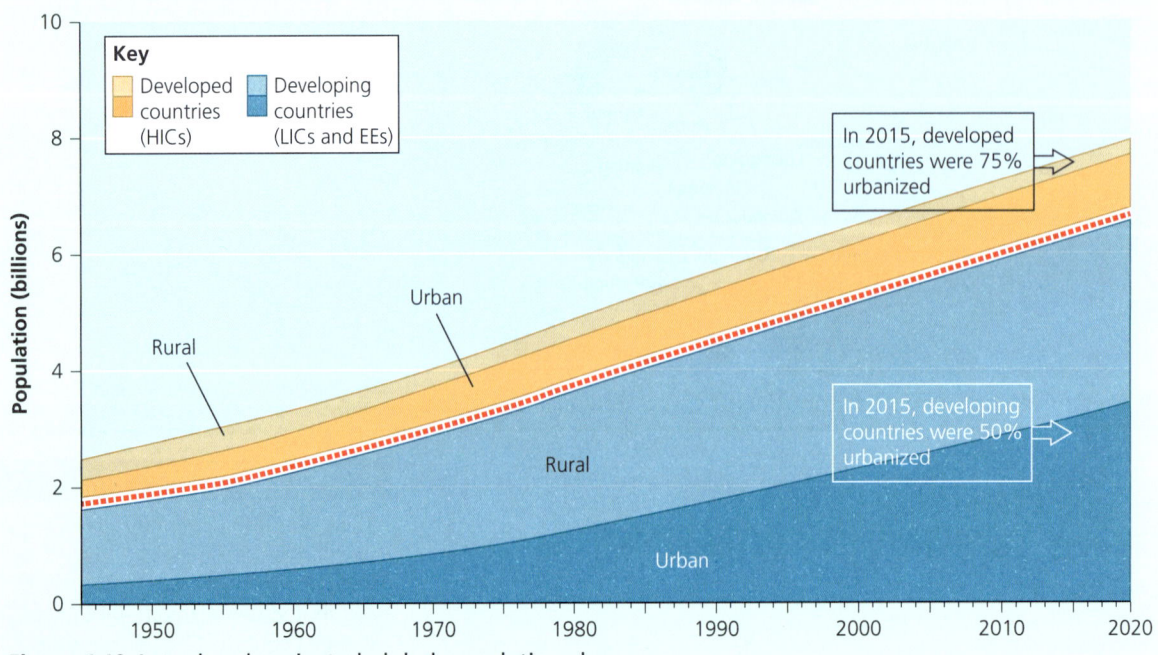

Figure 1.10 Actual and projected global population changes

in the 1980s and 1990s but slowed down after around 2010. Also shown are trends in urban growth for developed and developing countries. The number of people living in urban areas in developing countries has grown very rapidly and continues to rise steeply through the entire period shown. The number of people living in rural parts of developing countries stays the same after about 1980. There are far fewer people in either rural or urban parts of developed countries.

Examiner's comment

This is a poorly thought-out answer to the question set. The question does not ask for changes in total world population to be described. The first two sentences are therefore irrelevant. Trends for the developing world are described correctly although no supporting data are used. Trends for developed countries are not dealt with satisfactorily. Overall, this answer would only score 2 marks. Could you improve it?

Pay attention to the mark allocation

Although 4 marks are available overall, this question is shown to have a tariff of [2 + 2 marks]. Rather than answering with a single paragraph, it would be far better to write two short paragraphs. The first would deal with developed countries and the second with developing countries. By doing so, you would be able to judge more easily whether you have provided a balanced answer which deals equally well with developed and developing countries.

1.2 Changing populations and places

The study of population is also called **demography**. Geographers are interested in the demographic changes that occur in particular places over time. Processes of change vary greatly from place to place as a result of economic, cultural and political factors.

Keyword definition
Demography The study of population dynamics and changes.

Population change and demographic transition

Different countries may have very different demographic characteristics. These characteristics also change over time. Table 1.5 provides definitions of several key demographic terms and an overview of how they vary spatially and temporally.

Table 1.5 Key demographic indicators

Term	Definition	Spatial and temporal variations
Crude birth rate (CBR)	The number of live births per 1,000 people per year in a region.	• In pre-industrial societies, the rate is around 45/1,000/year: today, only a few remaining isolated rainforest tribes have a CBR close to this figure. The European average is 11/1,000/year. • The CBR becomes lower not only on account of falling fertility but also due to increasing life expectancy (which means a higher proportion of each 1,000 people are not of child-bearing age).
Crude death rate (CDR)	The number of deaths per 1,000 people per year in a region (a measure of mortality).	• In pre-industrial societies, the rate is 40–50/1,000/year. Today, the European average is 10/1,000/year. • The CDR can increase due to disasters or disease: in Haiti, in 2010, it rose from 8 to 32/1,000/year on account of a large earthquake.
Natural increase (NI)	The difference between the CBR and CDR. It can be recorded either as the net change per 1,000 people or as a percentage.	• It is rare for a country's rate of NI to exceed 3 per cent per annum. • Youthful migrant populations, such as those found in developing world megacities, have a high rate of NI due to the presence of a large proportion of fertile adults and relatively fewer older people.
Life expectancy	The average number of years a member of a particular society can expect to live.	• The average world life expectancy is now 71 (in 1960 it was 52). In most of Latin America and Asia, the figure is 70 years or higher. In 2015, the world's lowest was 49 in Swaziland; the highest was 84 in Japan (South Korea is predicted to overtake Japan soon, however). In most countries, national averages obscure a difference between men and women of around six years (Japan's female life expectancy is 87; for men it is 81).

Term	Definition	Spatial and temporal variations
Fertility rate	The average number of children a woman gives birth to during her lifetime.	• Today, most countries have an average fertility rate of three children or fewer, whereas in 1950 the world average was five. Fertility has fallen throughout Asia (Figure 1.11). • Some sub-Saharan African countries still have a high rate of seven or eight.
Infant mortality rate	The number of deaths of infants under one year old per 1,000 live births per year.	• Globally, estimates of the average rate vary between 42 and 49. This uncertainty reflects the difficulties of recording the data in an accurate and consistent way in different states. • Afghanistan had the world's highest rate in 2015 (115 per 1,000).

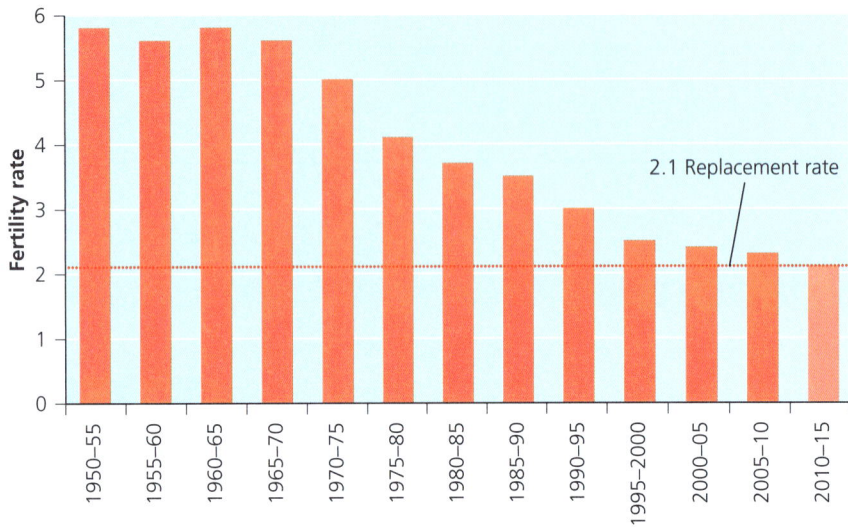

Sources: United Nations, World Bank Group

Figure 1.11 Falling fertility in Asia 1950–2014

Changing population structure and dependency ratios

The term **population structure** refers to the division of a population into a series of groups according to age and gender. The information is shown as a **population pyramid** for a particular historical year (Figure 1.12).

- People are divided into cohorts (age groups) placed either side of the vertical axis, with males on one side and females on the other.
- The size of each age group can be shown either as an actual number (in thousands or millions) or as a percentage of the total population.

This graphical expression of data allows us to see how fertility rates and life expectancy have affected a country or smaller-scale place's population structure. We may also be able to see what the effects of past migration, wars, disasters, economic recessions or epidemics have been on particular age groups, or men and women. Most European population pyramids for 2015 show a dip in the number of people aged 40 compared with those aged 45: this reflects a global economic crisis in the early 1970s which led many couples to delay having children. As a result, fewer babies were being born in 1975 (the people aged 40 in 2015) compared with 1970 (the people aged 45 in 2015).

When comparing the pyramids of different countries, important characteristics can be identified which offer clues about their relative levels of economic development.

- **Concave sides** to a country's pyramid can indicate low life expectancy and a very high death rate. This is because few individuals survive to move from one cohort to the next (but note that concave sides for a *city's* population pyramid might be caused by youthful in-migration rather than high mortality).
- A **wide base** for a country indicates a relatively high birth rate but low life expectancy. As a result, the pyramid narrows rapidly because large numbers of those born are not surviving to reach an older age (Figure 1.13).

> **Keyword definitions**
>
> **Population structure** The make-up of a population in terms of age, gender, occupation, ethnicity or any other selected criterion.
>
> **Population pyramid** A type of bar chart used to show the proportion of males and females belonging to different cohorts (age groups) for a place or country.

- **Perpendicular sides** relative to the base of a country's pyramid tell us that the majority of those born survive to an old age. This indicates a middle- or high-income county.

One common feature of most age–sex pyramids is more women than men surviving into the oldest cohorts. There are 110 women over the age of 60 for every 100 men in Europe.

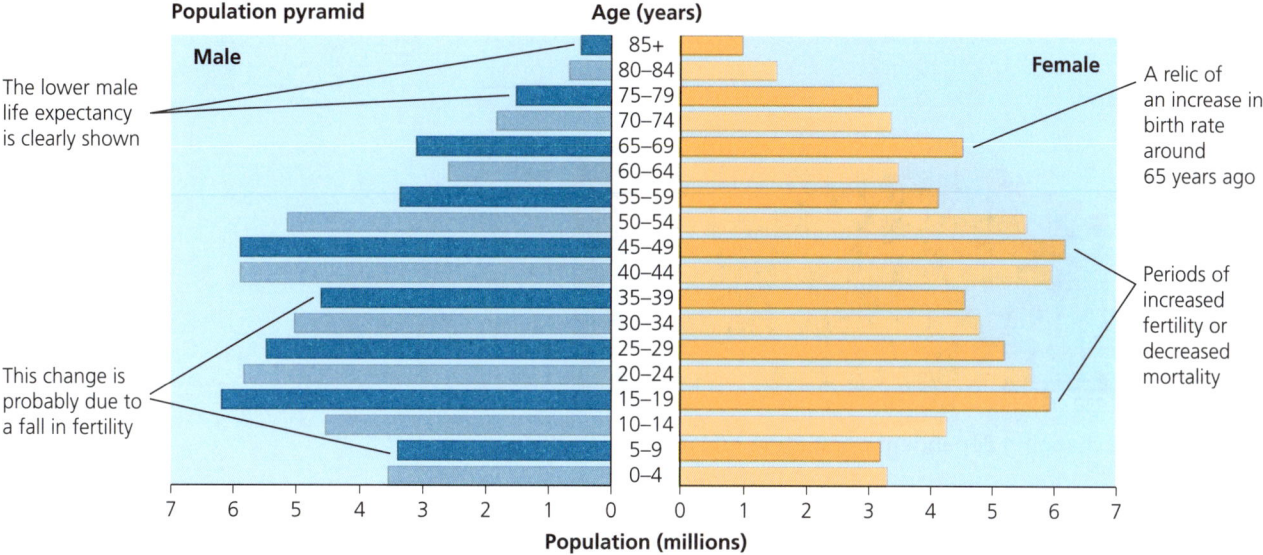

Figure 1.12 A population pyramid (the annotations help explain its irregularities)

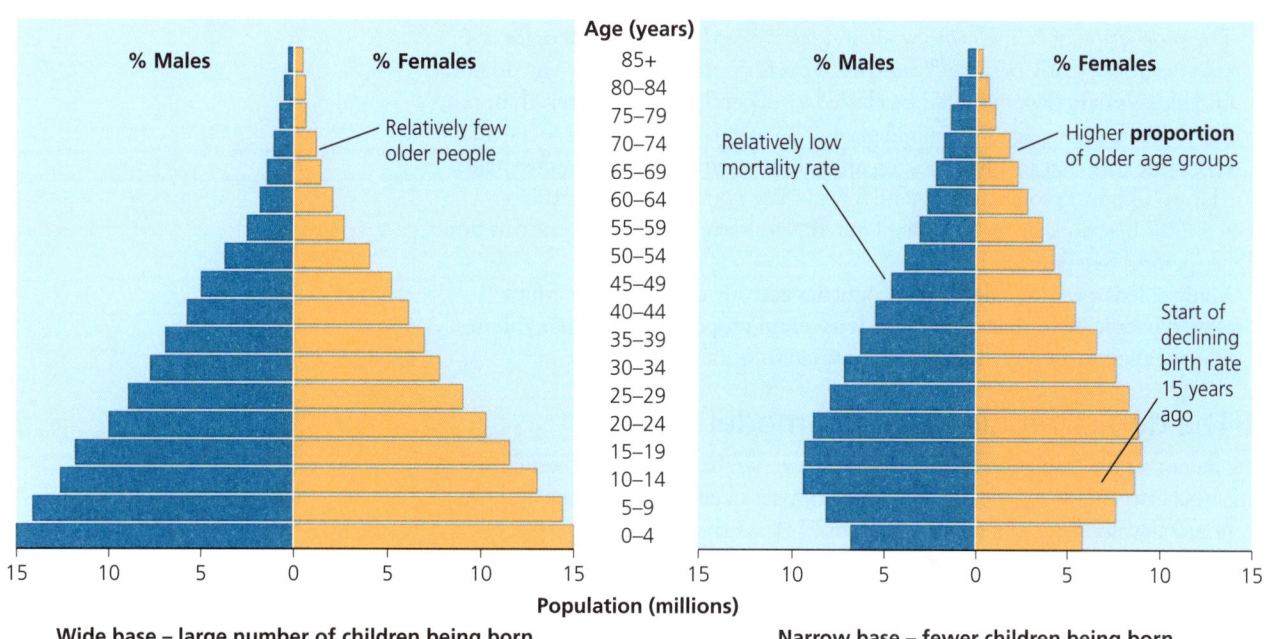

Figure 1.13 Contrasting population pyramids for a low-income and middle-income country

The dependency ratio

The dependency ratio is a shorthand measure that compares the proportion of a population that is economically 'non-productive' with the proportion that (in theory) generates wealth. In practical terms, the youngest and oldest cohorts are unproductive dependants whereas the middle-aged groups are wealth producers. The dependency ratio expresses the relationship between the active and inactive segments of a society using the following formula:

$$\text{dependency ratio} = \frac{(\% \text{ under } 15) + (\% \text{ over } 65)}{(\% \text{ aged } 15\text{–}64)} \times 100$$

A high score of around 60–70 suggests a lack of balance. It indicates that there are relatively high numbers of dependants in comparison with working taxpayers.

The dependency ratio is rising in many high-income countries and emerging economies as numbers of retired people increase. For instance, only 60 per cent of UK citizens are of working age. Taxes on their wages help pay for: the state pensions of the 21 per cent who are retired; the costs of schooling the young; and healthcare costs for both dependent groups. By 2030, only 56 per cent will be of working age. Dependency in Japan is predicted to become an even greater challenge (Figure 1.14).

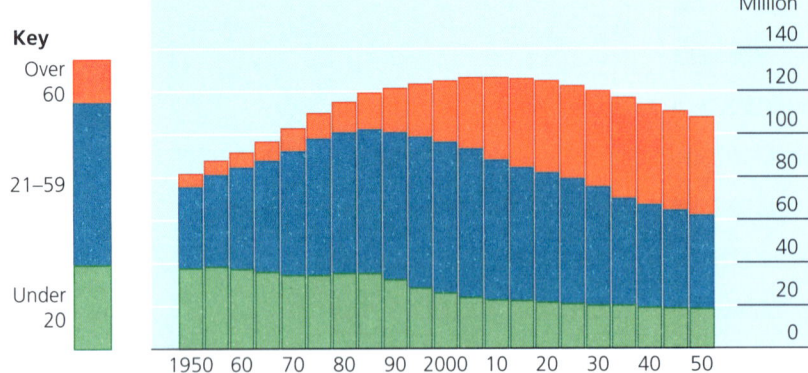

Figure 1.14 The changing dependent population of Japan, 1950–2050 (projected)

However, the dependency ratio is a very crude and potentially unreliable measure for several reasons:
- The age of 15 is a poor choice of dividing line for youthful dependency: in many countries, young people remain at school or college until later.
- The proportion of women who work and pay taxes that fund services for the elderly varies greatly between countries. This is not to say women who do not undertake salaried work should be classed as 'dependent'. More often than not, unsalaried working-age women are playing an essential (if unpaid) social and economic role that involves looking after children and managing households.
- Many children in some middle- and low-income countries do not attend school. They work instead, despite United Nations actions to prevent this from happening (see page 110).
- Many elderly people in developed countries actively create wealth. Many retired people own sizeable assets such as rental properties or company shares, which generate an income. Others continue to work (see page 27).

The demographic transition model (DTM)

The demographic transition model (DTM) is a generalized attempt to establish linkages between a sequence of population changes occurring over time and the economic development of a country (Figure 1.15). As the years pass, both vital rates (the CBR and CDR) are lowered. Table 1.6 shows possible reasons for this. Originally, the model distinguishes between four markedly different demographic phases:

1 **Traditional** ('high and fluctuating') stage: a pre-industrial agrarian society with no population growth.
2 **Early transitional** stage: the first phase of industrial development, which brings accelerating population growth.
3 **Late transitional** stage: a later phase of industrial development, which brings decelerating population growth.
4 **Advanced** ('low and fluctuating') stage: a 'post-industrial' or late capitalist society with little or no population growth in the longer term.

The model was based originally on data from England and Scandinavia. As such, it provides an insight into European demographic history and can be used as a starting point when speculating about what the future may hold for countries where population is still increasing due to a high rate of natural increase.

A fifth stage is sometimes added which shows the projected population decline for countries such as Japan.

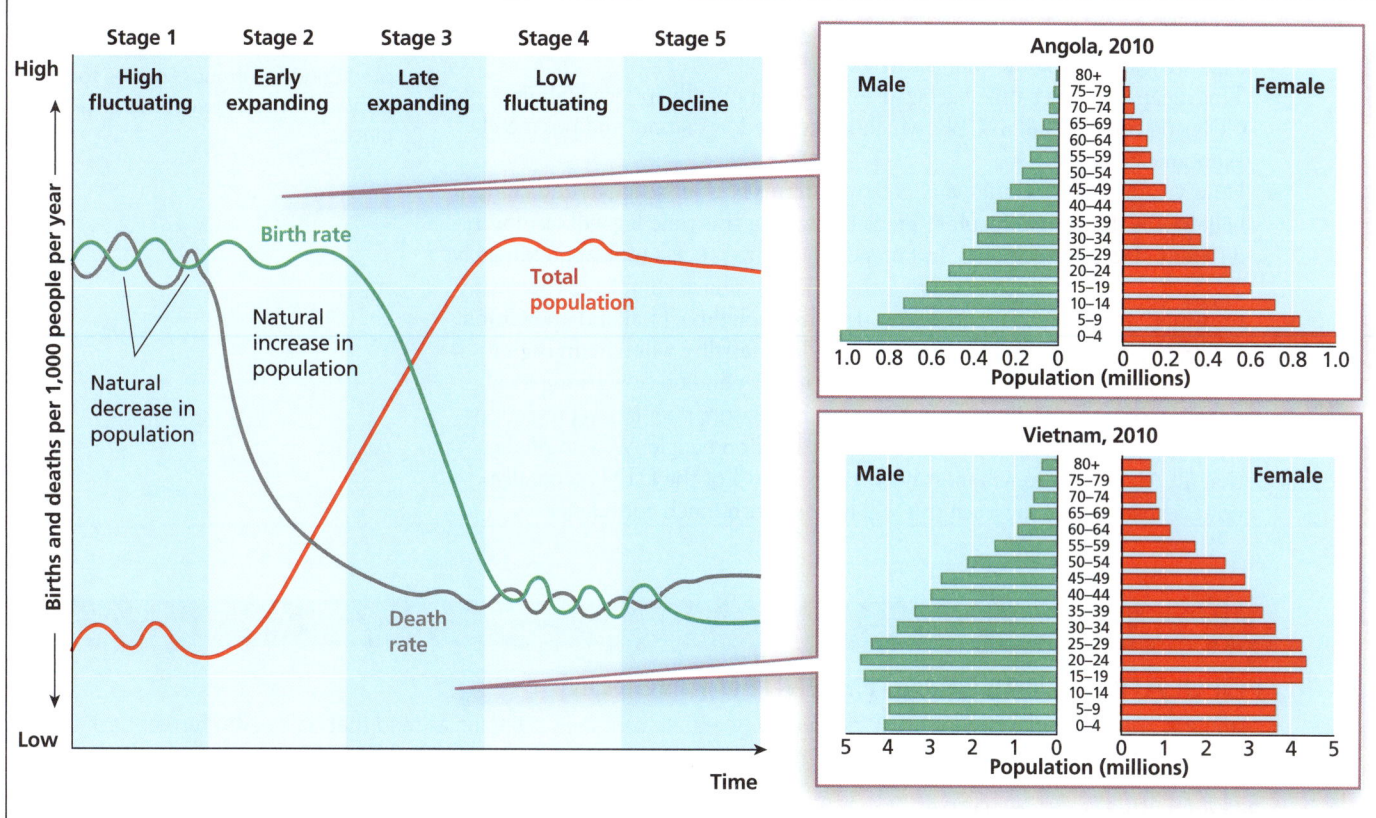

Figure 1.15 The demographic transition model (DTM)

Table 1.6 Reasons why the CBR and CDR fall over time as a country develops

Why has the CBR fallen over time in many places?	Why does the CDR fall when a country develops?
1 *Emancipation of women* Over time, women become more independent and choose to limit their family size, especially if they have career ambitions. 2 *Political changes* New laws increasing the age of marriage will reduce the number of births. 3 *Compulsory education* Well-educated people understand fully what the consequences of an unwanted pregnancy might be for their lives and careers; they are also aware of the availability of contraception. 4 *State healthcare* Government programmes may try to reduce teenage pregnancy and disease transmission by encouraging use of contraception. 5 *Secularization* In some countries, a weakening of religious beliefs has led to changing attitudes towards abortion and birth control. 6 *Materialistic society* In a consumer society, the cost of raising a child grows because of the expense of purchasing clothing, toys, holidays and technology. 7 *State welfare* In many countries, people no longer rely on children to care for them in old age because the state provides healthcare and pensions. 8 *Decline of the family as a unit of production* As countries develop, family life is no longer organized around subsistence farming. It may therefore make less economic sense to have many children.	1 *Improved food supply* Agricultural revolutions may include (1) the intensification of farming and increased food yields on existing lands and (2) the extensification of production to make use of remoter areas as transport infrastructure improves. For instance, between 1650 and 1750, important changes were made to crop rotation systems in Europe, allowing additional food to be grown. 2 *Healthcare* Disease prevention and treatments have improved globally. Well-planned economies can provide comprehensive care (e.g. the UK's National Health Service). Emergency services/ambulances can provide immediate help for injured people. Immunization against polio has led to its eradication in many states; smallpox has vanished. 3 *Hygiene, sanitation* and *safety* When sewer systems were installed in nineteenth-century European cities, life expectancy rose. Globally, primary school education aims to raise public awareness of the means by which diseases are transmitted. Most countries have health and safety laws that regulate many aspects of everyday life, including: road safety and seat-belt use; rules about smoking and alcohol use; fire exits for buildings; safety tests for technology and toys. We belong to risk-averse societies.

Unit 1 Changing population

Here are some important points to remember about the DTM:
- The crude death rate falls first (as food supply improves and medicine becomes available). Only much later does the crude birth rate fall (when women finally begin to produce fewer children as a result of more progressive cultural, legislative and economic changes in society).
- This means there is a time lag between the economically induced fall in CDR and the later fall in CBR, which is governed by *cultural* and not merely economic changes.
- However, different societies vary greatly in terms of their religious and cultural beliefs. It is therefore difficult to predict how long this time lag will last for in different places and societies. In Europe, the entire process took approximately 200 years to complete. In contrast, other countries have completed the entire transition more recently in just a few decades, including Thailand and China. Currently, there is no way of knowing how long fertility will remain high in many African countries and regions. That will depend on how strong local resistance to cultural change remains. For this reason, population projections for Africa in 2100 range from 2.5 billion to 4 billion people. We can predict that all African states will most likely enter stage 3 of the DTM *eventually*; however, we cannot say *when* this will happen with much confidence.

> **PPPPSS CONCEPTS**
>
> Think critically about the strengths and weaknesses of the demographic transition model as a way of predicting future possibilities for population change in different places.

CASE STUDY

POPULATION CHANGE IN THE UNITED KINGDOM

Birth rates and death rates fell in the UK during the twentieth century, continuing a trend that dates even further back to the early 1700s (Figure 1.16). Two important phases of population change stand out during the 1900s:

Phase 1: Before the 1970s

Population was still growing due to natural increase. The death rate had been falling since the 1800s, thanks to major improvements in food supply, health and hygiene. This trend continued during the early decades of the 1900s, bringing the death rate to its current low level by mid-century. However, it took until the 1970s, and the end of a post-Second World War 'baby boom', for the declining birth rate to reach the same low level as the death rate, finally bringing natural increase largely to a halt. To summarize, total population grew from 38 million to 55 million between 1901 and 1971, on account of a positive, though generally declining, rate of natural increase. (Note that low growth in the 1930s means that the UK is widely viewed as having entered stage 4 then, rather than in the 1970s.)

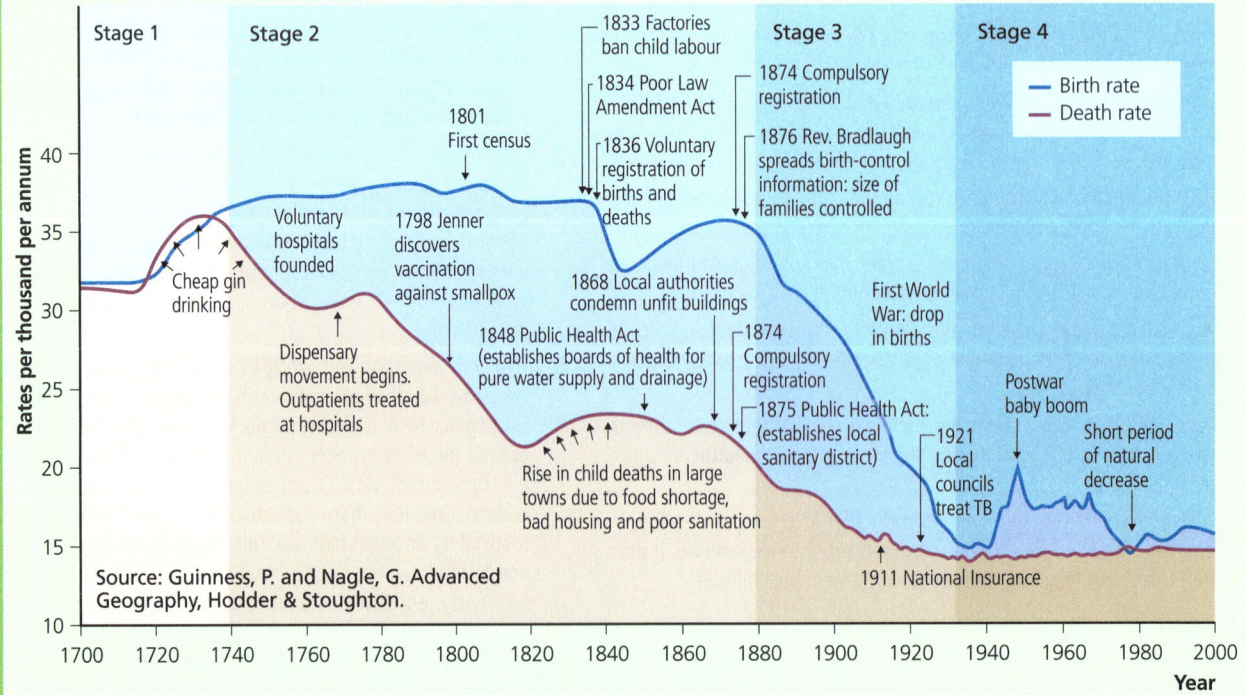

Figure 1.16 Changes in the CBR and CDR for the UK, 1700–present

Phase 2: Since the 1970s

Total population increased more slowly in the 1980s and 1990s. In recent years, new growth has mostly been due to immigration and people living longer (the number of over-80s will double from 3 to 6 million between 2010 and 2030). Both the death rate and the birth rate have remained at a low and constant level. As a result, family sizes are small while life expectancy keeps creeping ever upwards, thanks to science and medicine. However, the birth rate began to creep upwards again a few years ago (Figure 1.17). This is partly on account of the very large numbers of migrant women of child-bearing age currently living in the UK (a third of all new births are registered to foreign nationals). Also, more older women who have postponed having children are now managing successfully to have children at a later age, thanks to improved fertility treatments. To summarize, total population grew from 55 million to 65 million between 1971 and 2017, with migration and migrant births playing an increasingly important role.

Regionally, there are noticeable variations in the UK's population structure, fertility rate and life expectancy. The highest life expectancy is found in the Isle of Purbeck, where women now live on average to the age of 87. In contrast, male life expectancy in the city of Glasgow is 72.

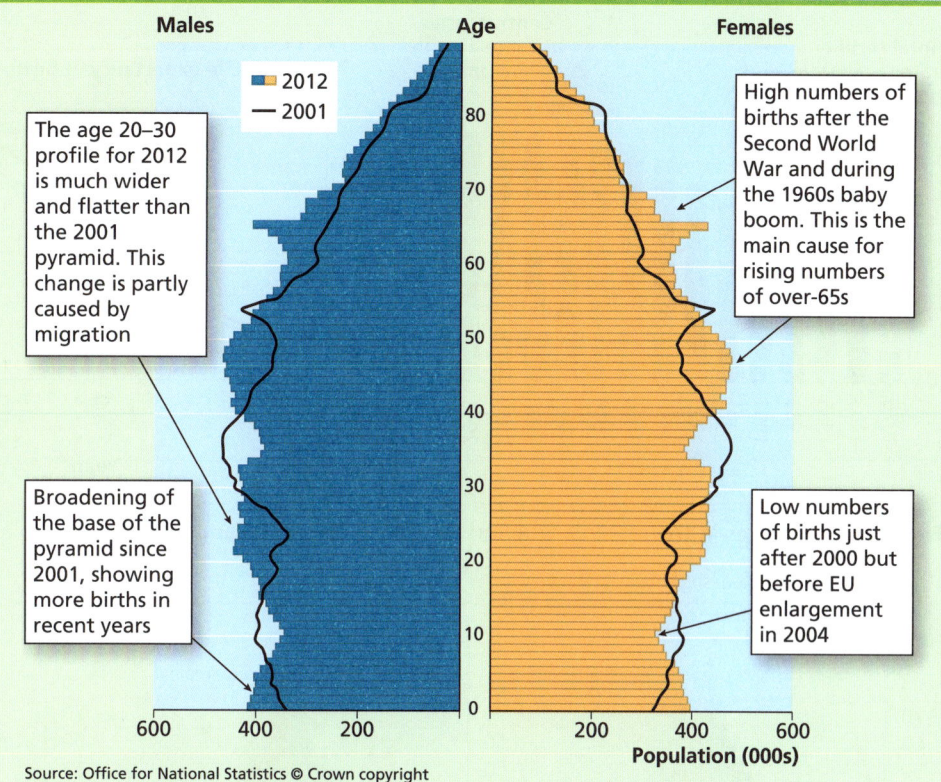

Figure 1.17 The UK's population pyramid, 2001 (outline) and 2012

Source: Office for National Statistics © Crown copyright

CASE STUDY

POPULATION CHANGE IN BANGLADESH

Bangladesh is well on the way to completing its demographic transition: the country is very close to entering stage 4 of the DTM (Figure 1.15). Between 1990 and 2010, life expectancy rose by 10 years, from 59 to 69 (the UK reached this figure in 1949). In part, this is due to improved healthcare and reduced infant mortality. Also, the famines which some Asian countries once suffered from have become a thing of the past, thanks largely to modern agricultural techniques and the **green revolution**.

The most striking demographic trend has been the way fertility has plummeted in just 30 years. Since independence in 1971, Bangladesh's government has adopted policies that have helped bring this about without coercion.

- Government employees and volunteers have worked tirelessly to distribute free contraceptive pills and advice across the entire country: in 1975, 8 per cent of women of child-bearing age were using contraception (or had partners who were); in 2010 the number was over 60 per cent (Figure 1.18).
- The proportion of girls who are schooled has increased over the same time period.
- Family planning, education and changing social attitudes have empowered women by giving them greater autonomy. As a result, more women have decided to follow careers and have delayed having children.
- The result is a fertility rate that fell rapidly from 6.3 in 1975 to 3.4 in 1993. Since then it has declined more slowly to reach 2.3 today. This is only slightly above the **replacement level** needed for a population level to stabilize in the long term.

> ### Keyword definitions
> **Green revolution** A period when the productivity of global agriculture increased greatly as a result of new technologies including fertilizers and selectively bred high-yield crops.
>
> **Replacement level** The fertility rate required to maintain a population at its current size.

18 Unit 1 Changing population

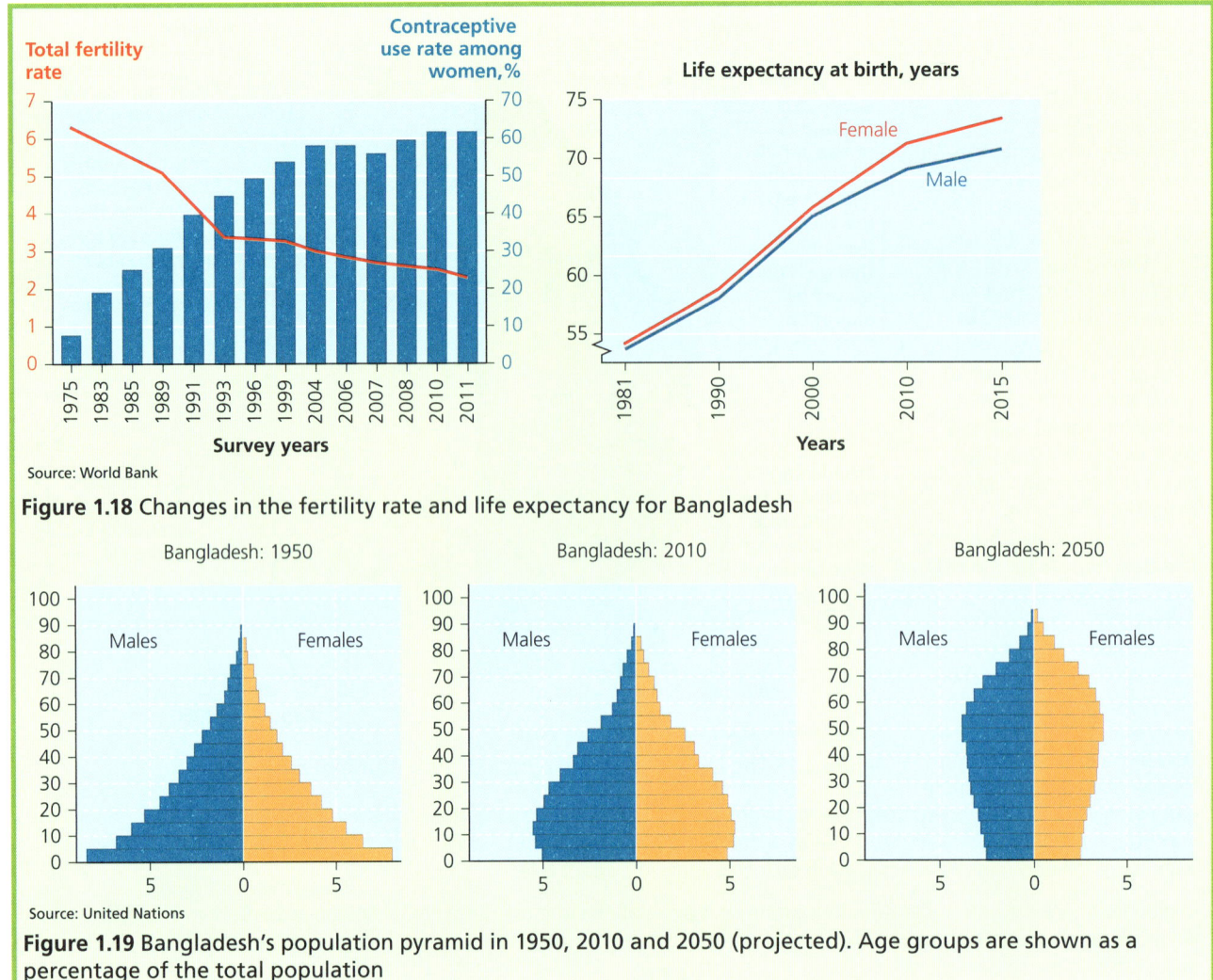

Figure 1.18 Changes in the fertility rate and life expectancy for Bangladesh

Figure 1.19 Bangladesh's population pyramid in 1950, 2010 and 2050 (projected). Age groups are shown as a percentage of the total population

Consequences of megacity growth

Revised

Urban growth on a very large scale can help provide a wide range of work opportunities for new migrants and established megacity residents. However, it may give rise to potentially insurmountable social and environmental challenges too.

PPPPSS CONCEPTS
Think about the social and economic changes that have empowered women in Bangladesh by giving them greater control over the number of children they would like to have.

Megacity challenges

Table 1.7 shows how two developing world megacities – Mumbai and Karachi – have grown over time. Continued growth of these and other megacities is inevitable. As wealth grows in developing countries and emerging economies, more young rural folk will develop aspirations beyond agriculture. Continued arrivals of large numbers of new people threaten the sustainability of these places, however.

Table 1.7 Examples of rapid megacity growth driven by rural–urban migration

Mumbai	Karachi
• India's Mumbai urban area is now home to 21 million people, having more than doubled in size since 1970. People flock there from the impoverished rural states of Uttar Pradesh and Bihar. • Urban employment covers a range of sectors and skill levels. Global brands like Hilton and Starbucks are present in Mumbai. In retail areas, like Colaba Causeway, large numbers of people work selling goods to the country's rising middle class. • Dharavi is a slum housing area in Mumbai. It has a buoyant economy: 5,000 people are employed in its plastics recycling industries. However, rising land prices across Mumbai mean there is great pressure to redevelop this and other slum areas.	• Before Islamabad was founded in 1960, the port city of Karachi was the capital city of Pakistan. • Approximately 24 million people lived in Karachi in 2015, making it the most highly populated city in Pakistan and one of the world's five most populous megacities (Tokyo is the largest). • This colossal megacity is Pakistan's centre of finance, industry and trade. People flock to the city for work from rural areas all over Pakistan, including the Sindh and Punjab provinces. • Once living there, rural–urban migrants can find formal or informal employment in a range of industrial sectors including shipping, banking, retailing and manufacturing.

1.2 Changing populations and places

Figure 1.20 shows three interrelated goals that city managers must pursue if they are to provide for sustainable urban living. Top-down strategies that can help deliver one or all of these goals include robust policies for job creation, such as the establishment of **Special Economic Zones (SEZs)** where low tax rates are used to attract foreign investors and promote industrial growth. SEZs in Jakarta (Indonesia) and Shenzhen (China) have made these megacities magnets for investment and migration (Figure 1.21). However, they have also given rise to 'overheating' problems, including traffic congestion and air pollution (dubbed 'airpocalypse' in China).

China has accommodated rural–urban migration by planning and constructing new housing and cities on an enormous scale. Three major megacity clusters now exist in China's Yangtze River Delta (including Shanghai), Pearl River Delta (including Shenzhen, which used to be just a fishing village) and the Bohai Sea rim (including Beijing).

However, even the best-planned housing policies can struggle to match demand with supply in megacities where hundreds of thousands of new arrivals are added to the population annually. Housing shortages are especially acute in the African megacities of Lagos and Kinshasa, where many people live in slums.

Keyword definition

Special Economic Zone (SEZ)
A part of a city or country where business tax and trading laws are more liberal than those found in the rest of the state, for the purposes of stimulating investment and industrial activity.

Figure 1.20 Three interconnected goals for sustainable megacity management

Figure 1.21 Jakarta, Indonesia

Bottom-up urban community development strategies may also play an important role in megacity management.

Poor migrant communities in Lagos, Nigeria must take 'bottom-up' steps to improve their local environment and access to housing by themselves without much state support. As a result, squatter settlements have grown throughout Lagos. They are densely populated owing to the shortage of available housing and land.

In the case of Makoko, a slum settlement on the edge of Lagos Lagoon, makeshift homes have been built above the water on stilts. People use materials like tin sheets and wooden planks. They have also reclaimed land from the lagoon using waste materials and sawdust to create new islands for building on. The population of Makoko is estimated at one quarter of a million people. Most people make a living in the informal economy, including fishing. This goes back to Makoko's origins as a fishing village outside Lagos: as the city grew it has been swallowed up in the urban area.

> **CASE STUDY**
>
> ## DHAKA
>
> According to the World Bank, Dhaka in Bangladesh is the fastest-growing megacity in the world. Between 1990 and 2005, the city doubled in size from 6 to 12 million. In 2015 its population reached 15 million and by 2025 Dhaka will be home to an estimated 20 million people (making it larger than either Mexico City or Beijing). The challenges for Dhaka are extreme, even by the standards of other developing world megacities. The Bangladesh Centre for Advanced Studies estimates that half of Dhaka's population lives in slums, many of which have grown in unsafe areas beside rail lines, along riverbanks and in swampy and disease-ridden areas (Figure 1.22).
>
> - According to one report titled *The Megacity of the Poor*, about 70 per cent of Dhaka's households earn less than US$170 per month; many take home less than half that.
> - Many migrants arriving in Dhaka will work for low pay in the country's textile industry, which is its largest economic sector. Foreign TNCs have flocked to Dhaka in recent years. It is likely that you and your friends will own clothes made in Bangladesh.
> - There has been international concern over standards of health and safety at work in Bangladesh following the highly publicized collapse of the Rana Plaza factory in 2011, which led to the deaths of over one thousand workers.
>
> Yet rural migrants continue to flood into Dhaka. They believe – rightly or wrongly – that life there will be an improvement on the village they have left behind.
>
> - The Bhola district, for instance, is located in southern Bangladesh. This region is vulnerable to climate change.
> - Migration is seen as a potential adaptation strategy to the heightened risk of flooding; some people have abandoned the area altogether. In other cases, families are attempting to relocate to higher or better-drained land. To meet the costs of moving, it becomes necessary for some household members to migrate to urban areas. There, they work and send money home as remittances to help their family adapt.
> - Women migrate in especially large numbers: a distinct Bhola slum has developed in Dhaka. It is home to thousands of female garment factory workers who have made the move there from Bhola district.
>
> Dhaka's government has plenty of issues to be concerned with in addition to population and slum growth.
>
> - Recently, the city has been a target for terrorists (in 2016, 20 people were shot dead in a restaurant).
> - The city is also very vulnerable to flooding during the monsoon season (April–July). The 1988, 1998 and 2004 floods were particularly severe and brought large economic losses.
>
> Yet despite these challenges, Dhaka is also the crucial growth engine for an emerging economy whose per capita GDP has risen from US$300 in 1970 to over US$1,400 today. When adjusted for PPP (see page 5), the current figure is actually closer to US$4,000.
>
>
>
> **Figure 1.22** A Dhaka slum
>
> **PPPPSS CONCEPTS**
>
> Think critically about the possible costs and benefits of relocating to Dhaka for a teenage girl living in poverty in rural Bangladesh.

Forced migration, refugees and internally displaced people

Revised

Refugees are people who have been forced to leave their country. They are defined and protected under international law, and must not be expelled or returned to situations where their life and freedom are at risk. In addition to refugees, many people worldwide have become **internally displaced people (IDPs)** after fleeing their homes. In 2016, the conflict in Syria, which started in 2011, had generated 5 million refugees and 6 million IDPs; half of those affected were children.

According to United Nations data:
- In 2014, more people were forced to migrate than in any other year since the Second World War. Around 14 million people were driven from their homes by natural disasters and conflict. On average, 24 people were forced to flee their homes each minute – four times more than a decade earlier.

Keyword definitions

Refugee A person who has been forced to leave his or her country in order to escape war, persecution or natural disaster.

Internally displaced people (IDPs) People who have found shelter in another part of their country after being forced to flee their homes.

- The global total of displaced people now exceeds 60 million. Of these, around 40 million are internally displaced and 20 million are refugees.
- Recent forced movements of people have been caused by wars in Syria, South Sudan, Yemen, Burundi, Ukraine and Central African Republic. Thousands more have fled violence in Central America. Figure 1.23 shows the list of countries that have generated most refugees on a rolling annual basis since the 1970s.

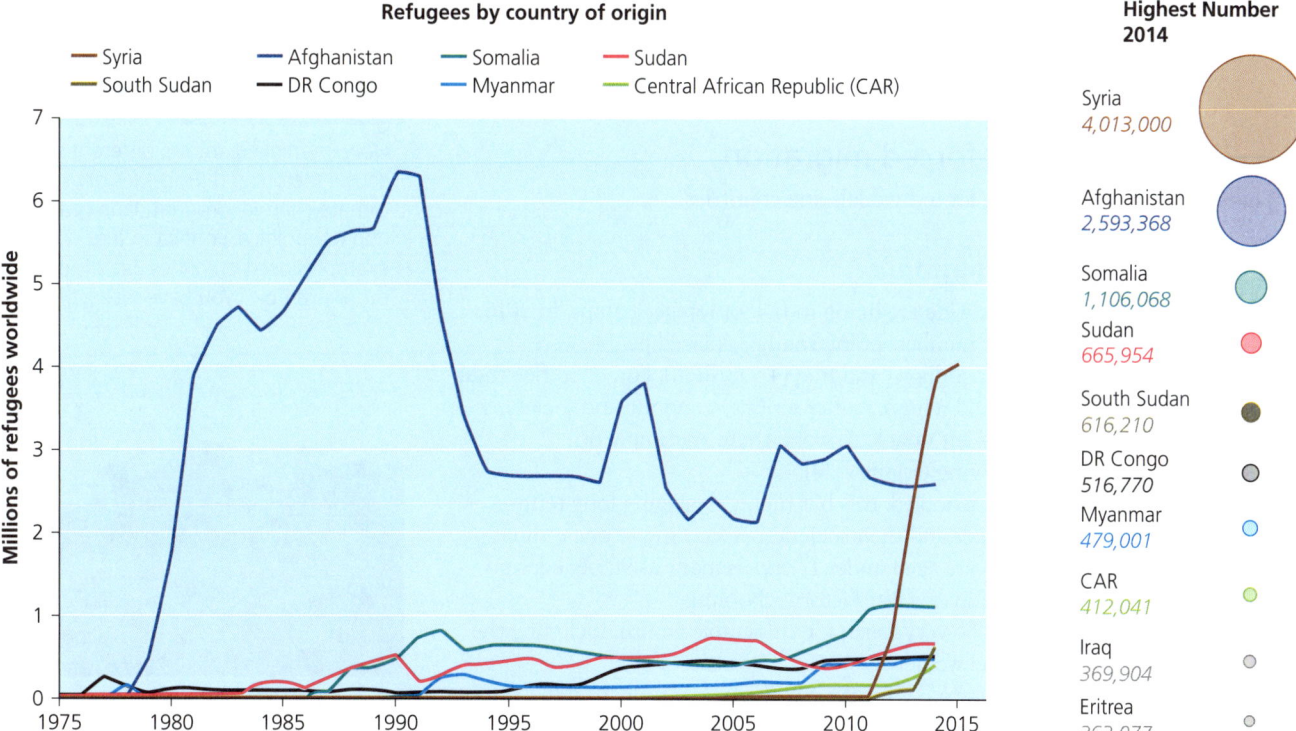

Figure 1.23 The main source countries for refugees 1975–2016 (fluctuations indicate periods when many people fled each state and also years when large numbers returned home)

Causes of forced migration

There are many reasons why people are forced sometimes to flee their homes. These include conflict, land grabbing and climate change.

Conflict

Between 1945 and 1970, most previously colonized African countries finally gained their freedom from European rule and became independent states. In the decades since then, widespread geopolitical instability within many of these newly independent countries has led to the widespread forced migration of people. Today, there are over 2 million refugees scattered across Africa. Large numbers are living in Chad, Kenya and Sudan; significant source countries include Somalia, Rwanda, Democratic Republic of the Congo (DRC) and Angola.

Land grabbing

This is an economic injustice that involves the acquisition of large areas of land in developing countries by domestic governments and individuals, and MNCs. In some instances, land is simply seized from vulnerable groups by powerful forces and not paid for. Indigenous groups, such as subsistence farming communities, may have no actual legal claim to their ancestral land. They sometimes lack the literacy and education needed to defend their rights in a court of law. Around the world there are many instances of unjust land grabs resulting in social displacements and refugee flows. You may be familiar already with the example of Amazonian rainforest tribes losing their land to logging companies.

■ Climate change

Climate change acts to intensify rural poverty and conflict in some countries. Movers who might previously have been classed as economic migrants become refugees due to an increasingly hostile environment.
- Sudan's semi-arid Darfur region is home to black African farmers and nomadic Arab groups. Between 2003 and 2005, land grabbing and conflict led to the displacement of 2 million people. In this case, competition over land was exacerbated by drought, **desertification** and shrinking water supplies.
- Since 1990, millions of refugees have moved to escape drought in the Horn of Africa; many have moved from Somalia and Ethiopia into neighbouring Kenya.

Consequences of forced migration

There are consequences both for individuals and countries.

■ Consequences for migrants

Many millions of people worldwide are living in IDP or refugee camps. In 2016, the countries with the largest numbers of internally displaced people were Colombia (7 million), Syria (7 million) and Iraq (4.5 million). Forced to flee their home and possessions, IDPs and refugees suffer serious economic and social losses.
- In many camps, adults are left unable to work: there are simply no opportunities to make a living (Figure 1.24).
- Children may cease to be schooled: this has highly damaging long-term impacts for these individuals and their societies. According to one estimate, half of all forced migrants are aged under 17 and as many as 90 per cent no longer receive any education or a satisfactory schooling.
- Life in refugee camps can be very tough for vulnerable groups, including the elderly, the very young and women.
- Many refugees who have escaped horrific conditions continue to suffer trauma long afterwards as a result. This includes large numbers of people who were forced to fight as child soldiers in Sierra Leone and DRC in the 1990s.

■ Consequences for neighbour states

The majority of refugees do not attempt an ambitious long journey to a distant developed country. Instead, they travel no further than the nearest state neighbouring their home country. For families with young children and sick, injured or elderly relatives, it is easy to see why this is the case. This puts enormous pressure on states with the largest numbers of refugees, such as Turkey (where over 2 million Syrians now live).

■ Consequences for high-income countries

Since 2006, rising numbers of refugees from North Africa and the Middle East have attempted to reach Europe by crossing the Mediterranean in unsafe fishing boats. By 2016, an estimated one million people had attempted the crossing, including many refugees of varying faiths and ethnicities from Syria and poor, war-torn African nations like Somalia, Eritrea and Ethiopia. More refugees have walked all the way to Europe from Syria, arriving in large numbers at the borders of Hungary and Serbia in 2015 and 2016. Many issues arise for European countries as a result of this mass movement of people:
- EU coastguards have struggled to prevent deaths at sea in the Mediterranean. Around 800 people died when a boat capsized in rough seas off the Italian coast near Lampedusa in April 2015. By the end of the same year, around 3,700 people had died in similar circumstances. A further 160,000 people were rescued at sea.
- All EU states – along with most other countries – are obliged to take in refugees, irrespective of whatever economic migration rules exist. This is because they have signed the Universal Declaration of Human Rights (UDHR),

> **Keyword definition**
> **Desertification** The intensification or extensification of arid, desert-like conditions.

> **PPPPSS CONCEPTS**
> How interlinked are the different causal factors and processes of climate change, drought, land grabs and geopolitical conflict in the examples used on pages 22–24, or other examples you have researched?

Figure 1.24 An IDP camp in Sulaymaniyah, Iraqi Kurdistan

which guarantees all genuine refugees the right to seek and enjoy asylum from persecution (Table 1.8). The cost of care can be considerable, however. It is estimated to cost US$20,000 a year to cater for the needs of one recently arrived refugee who may be recovering from severe physical or psychological harm.
- Although the number of refuges granted asylum in 2015 amounted to less than 0.1 per cent of the EU's population, many European citizens are unhappy with what they view as a high number. It has become an emotive and increasingly divisive issue that divides communities and affects people's voting behaviour in elections. The debate intensified after a suicide bomber in the Paris attacks of December 2015 was identified as a Syrian refugee who had travelled to France via Greece.
- However, in the longer term many forced migrants will find employment and will contribute to the economy of the states that have granted them refuge.

Table 1.8 How the United Nations offers protection to refugees

The Refugee Convention (1951) and Convention relating to Stateless Persons (1954)	• The 1951 Refugee Convention is the key legal document that forms the basis of all UN work in support of refugees. Signed by 144 states, it defines the term 'refugee' and outlines the rights of refugees, as well as the legal obligations of states to protect them. The core principle is 'non-refoulement'. This means that refugees should not be returned to a country where they face serious threats to their life or freedom. This is now a core rule of international law. • The 1954 Convention relating to Stateless Persons was designed to ensure that stateless people enjoy a minimum set of human rights. It established human rights and minimum standards of treatment for stateless people, including the right to education, employment and housing.
The Office of the United Nations High Commissioner for Refugees (UNHCR)	• UNHCR serves as the 'guardian' of the 1951 Refugee Convention and other associated international laws and agreements. It has a mandate to protect refugees, stateless people and people displaced internally. On a daily basis it helps millions of people worldwide at a cost of around US$5 billion annually. UNHCR works often with the UN's World Health Organization (WHO) to provide camps, shelter, food and medicine to people who have fled conflict. • UNHCR also monitors compliance with the international refugee system.

CASE STUDY

MILITIA GROUPS

Since 2000, destabilizing **militia groups** have arisen in numerous local contexts, including al-Qaeda in the Arabian Peninsula and the Taliban in Afghanistan. Daesh (or IS) wages its so-called jihad against all other religions. Its soldiers have pursued a strategy of annihilating minority communities including Christian Assyrians, Kurds, Shabaks, Turkmens and Yazidis. Many people have become refugees.

> **Keyword definition**
>
> **Militia groups** An armed non-official or informal military force raised by members of civil society. Militia groups may be characterized as either freedom fighters or terrorists in varying political contexts, or by different observers.

The Boko Haram militia group has gained notoriety on account of its brutal actions in Borno state in Nigeria, and in neighbouring Niger. 'Boko Haram' translates loosely as 'Western education is forbidden'. The militants are strongly opposed to what they see as the spread of Western culture, among which they include the culture of gender equality.

- In 2014, 200 Nigerian schoolgirls were abducted by the Boko Haram militant group. Other targets have included the police, students, the media, churchgoers and ordinary civilians.
- By 2016, 1.4 million people had been displaced throughout Borno state owing to Boko Haram, while in Niger's Diffa region more than 280,000 people have been forced from their homes. In total, Boko Haram may be responsible for 2.5 million IDPs.

Life remains tough for people even when they have found shelter in United Nations IDP camps. Human Rights Watch is a civil society organization (CSO) that has documented the rape of women and girls living in IDP camps in Maiduguri, in northeast Nigeria. Camp dwellers also live in constant fear of further attacks from Boko Haram.

You can find out more about the humanitarian crisis caused by Boko Haram at: http://www.unocha.org/nigeria/about-ocha-nigeria/about-crisis

CASE STUDY

THE POLITICAL AND ENVIRONMENTAL CAUSES OF SYRIA'S REFUGEE CRISIS

In recent years, the Middle East has become the world's largest source region for refugees. Like central Africa, there is an unhappy fit between state borders and the Middle East's ethnic, cultural and religious map. The Sykes-Picot line was drawn by the British and French in 1916. It split apart large Sunni and Shia Muslim communities and led to the creation of several inherently unstable states including Iraq and Syria. The BBC calls it 'the map that spawned a century of resentment'.

The current crisis in Syria began when rebel groups demanded the resignation of Syria's ruling President Bashar al-Assad in 2011. The EU and USA initially supported some rebels but by 2015 found themselves bombing Daesh in Syria, effectively acting alongside Assad's forces. Russia and Saudi Arabia have provided funding for rival armies of groups, fuelling the conflict further. Figure 1.25 shows the enormous numbers of refugees fleeing Syria since the crisis began. You can see how the majority have travelled to Turkey, Lebanon and Jordan rather than EU states.

Environmental factors

US security analysts based at the Pentagon have in part attributed Syria's refugee crisis to desertification. From 2006 to 2011, large areas of Syria suffered an extreme drought that, according to scientists, was exacerbated by climate change (see page 22). The drought lead to increased poverty and relocation to urban areas. In turn, rising unemployment in cities triggered unrest and conflict.

PPPPSS CONCEPTS

Explain two strengths of the way data showing interactions between places have been represented graphically in Figure 1.25.

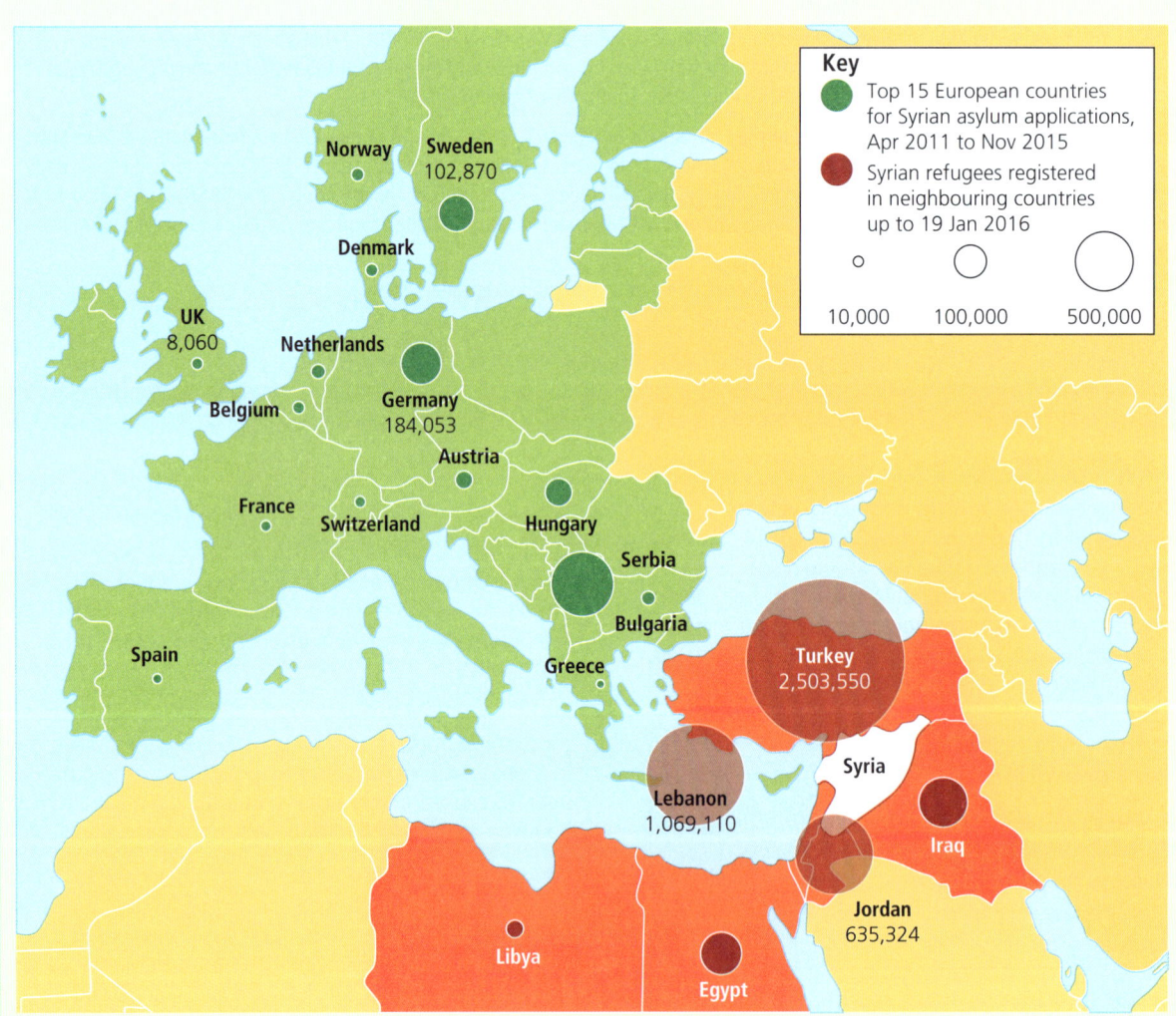

Figure 1.25 The distribution pattern of Syrian refugees, 2015

1.3 Population challenges and opportunities **25**

■ KNOWLEDGE CHECKLIST:

- The main components of population change, including the CBR and CDR
- Reasons for variations in population structure and dependency ratios
- The demographic transition model (DTM)
- Detailed examples of population change (the UK and Bangladesh)
- The challenges of megacity growth for societies, including the need for top-down planning
- The challenges of megacity life experienced by individuals in poor communities
- A case study of megacity growth (Dhaka)
- Global trends in refugee flows and movements of internally displaced people (IDPs)
- The human and physical causes of forced migration
- The consequences of forced population movements for migrants, neighbouring countries and high-income countries
- Two contrasting examples of forced movements with political and environmental causes (Syria's refugee crisis and militia groups, including Boko Haram)

EVALUATION, SYNTHESIS AND SKILLS (ESK) SUMMARY:

- How some population changes and movements involve spatial interactions that link different places together
- How population changes, interactions and movements are represented graphically using trend lines, charts and maps
- How push factors for refugees are interlinked

EXAM FOCUS

COMPLETE THE ANSWER

Below is a sample AO2 short-answer question, and an answer to this question. The answer contains a basic point and an explanatory link back to the question, but lacks any development or exemplification. Without this, full marks are unlikely to be awarded. Write out the example answer below, then improve the two basic reasons that have been given by adding your own development or exemplification.

Explain two possible negative consequences of megacity growth. [2 + 2 marks]

1. The first possible negative consequence is traffic congestion and air pollution. This is because major cities are home to such large numbers of people.
2. The second possible consequence is a lack of green space. The sheer number of people moving to megacities means that all open areas quickly become built on.

Answer short questions concisely

In both of the answer spaces, space has been used up in repeating the question ('The first possible negative consequence is…'). Try to avoid doing this.

ELIMINATE IRRELEVANCE

Below is a sample AO2 short-answer question and an answer to this question. Read the student answer carefully and identify parts that are not directly relevant to the question asked. Write out the answer, then draw a line through the information that is irrelevant and justify your deletions in the margin.

Suggest two possible economic reasons for a falling fertility rate in a country. [2 + 2 marks]

1. Economic changes can include industrialization and the decline of subsistence farming. When families no longer work together to farm the land, there is less economic benefit in having a large number of children. Fertility in Europe fell in the 1800s when large numbers of people abandoned subsistence farming and moved to cities to find paid employment.
2. Over time, important social changes often take place. For instance, fewer people may pay strict attention to religious teaching which prevents the use of contraception. Once this has happened, it is common to see the fertility rate falling in a country. Throughout Asia the fertility rate is now generally lower than three children per woman.

1.3 Population challenges and opportunities

Revised ☐

Global and regional population trends

Revised ☐

During the twentieth century, the world's high-income countries experienced significant changes in (1) average family size and (2) the proportion of people surviving to an older age, giving rise to the phenomenon of an **ageing population**. An important development in some low- and middle-income countries has been changing social attitudes towards the role of women in society. This has affected the **sex ratio** of some societies.

■ The global issue of ageing

In Japan, Iceland, Italy, Australia, Germany and many other countries, life expectancy is now 80 or higher. More than 20 per cent of the population in these countries are currently aged 65 or over. In the future, the proportion of elderly citizens in developed countries will grow higher still, while most middle-income countries will also begin to experience the effects of widespread ageing (Figure 1.26). Globally, the number of older persons aged over 60 is expected to exceed the number of children aged under 15 for the first time by around 2050, helped significantly by the long-term legacy of China's one-child rule (which left the world's largest nation of 1.3 billion people with its own greying 'time-bomb').

Migration can be another cause of an ageing population. When Poland joined the European Union in 2004, the total number of Polish emigrants living overseas rose quickly from 2 million to 3 million. Out-migration of the young has accelerated the rate at which Poland's population is ageing: loss of the young means that those who are not make up a larger proportion of a shrunken population.

> **Keyword definitions**
>
> **Ageing population** A population structure where the proportion of people aged 65 and over is high and rising. This is caused by increasing life expectancy and can be further exaggerated by the effect of low birth rates. It is also called a 'greying' population.
>
> **Sex ratio** The relative proportions of men and women in a society's population.

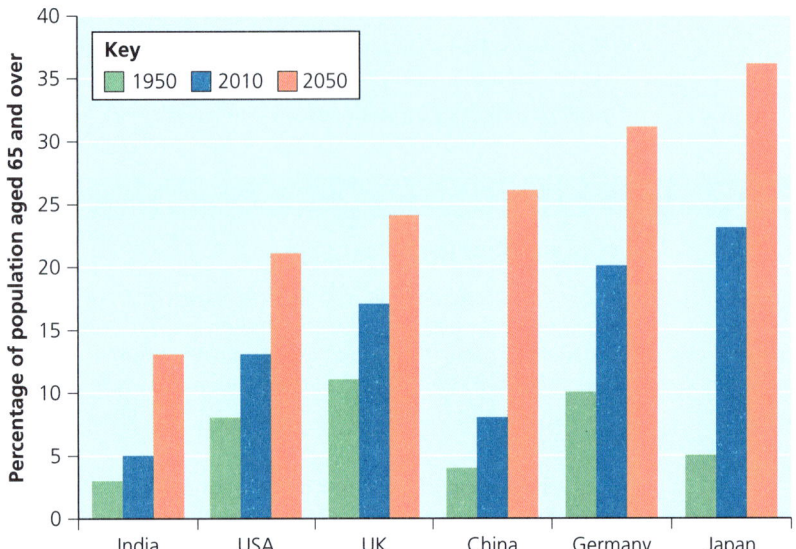

Figure 1.26 Actual and projected ageing population trends for selected countries

A key concern for governments of countries with an ageing population is meeting the economic cost of care for the growing proportion of people who – in many cases – no longer generate much wealth themselves. Potentially enormous amounts of money must be made available for the elderly's day-to-day living expenses, health treatments and housing costs. Japan's healthcare and nursing home costs by 2025 are expected to be almost US$1 trillion (about 12 per cent of GDP).

In additional to the financial costs of an ageing population, rising longevity is placing an emotional burden on younger and middle-aged people who must increasingly act as (unpaid) carers for older relatives who have developed Alzheimer's disease or another form of dementia. Thanks to improvements in healthcare, fewer people now die in their 60s and 70s from cancer, heart diseases or strokes. Instead, more people now survive into their 80s and 90s, ages at which vulnerability to degenerative brain diseases and disorders becomes greatly increased. Many young people are totally unprepared for the role reversal that will occur in later life when, instead of being cared for by their parents, the situation is reversed. Moving relations to a care home may help make life more manageable, but the effects of Alzheimer's are still upsetting to witness in a loved one.

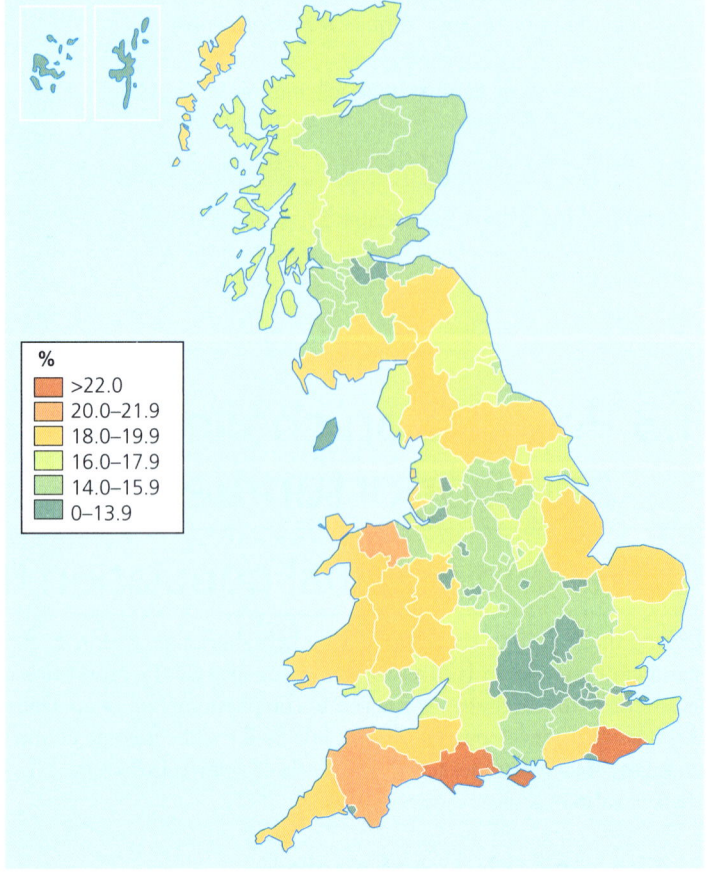

Figure 1.27 Regional variations in the proportion of the population aged 65 and over in England, Scotland and Wales

Ageing population challenges at the local scale

The issues are amplified further at the local scale sometimes. In countries where homes and services for the elderly are funded by local taxes levied on working people, districts with an unusually high proportion of elderly face the most severe financial challenges. In the UK, for instance, many coastal settlements have high proportions of older people living there as a result of age-selective migration (Figure 1.27).

The benefits an ageing population can bring

It is wrong to only view the elderly as a burden on society, however.
- Many elderly people who have worked in skilled occupations such as banking, teaching or medicine continue to work into their 70s and sometimes 80s. They may serve as consultants, providing valuable advice to younger staff, perhaps on a part-time basis.
- Older actors are always needed to play film parts – for example, Dumbledore and other older Hogwarts teaching staff in the Harry Potter films (Figure 1.28).
- In politics, the billionaire Donald Trump was elected as US President at the age of 70.
- Elderly people often make a range of contributions to society by assisting with the running of charities and good causes. Some assist their own families by providing free childcare for grandchildren, thereby allowing both parents to participate in the workforce.

It is important also to note that the very idea of an 'elderly population' aged over 65 is increasingly an over-simplification. There is usually a great difference – in terms of physical health, stamina and the ability to continue generating income – between highly active people in their 70s and far more infirm people aged 90 and over.

> **PPPPSS CONCEPTS**
>
> Think about the processes that give rise to unequal numbers of older people living in different places at varying scales.

Figure 1.28 Many people over retirement age contribute economically in highly significant ways, including actor Michael Gambon

Trends in family size

The global fall in fertility rates was explained in Table 1.6 (page 15). In most countries, women now give birth to fewer than three children on average. This has had the effect of reducing the average size of a family. In the past, it was usual to find two parents sharing their home with up to six or more children in European, Asian and South American countries. Today this is far less common and a typical nuclear family comprises between three and five individuals living under the same roof.

Other important social changes have also led to a reduction in the average family or household size.

- In many high-income countries, it is less common now to find older people living with their children and grandchildren. The extended family (where three generations live together) has become largely a thing of the past in some places. Government support in the form of pensions and healthcare has meant older people can remain independent and no longer need to rely on their children to support them in old age, thus, they may continue to live alone.
- In some high-income countries, including the UK, many older people own their houses outright because they bought them at a time when prices were lower. The cost of remaining there is therefore low. In Europe, an increasing number of elderly widows aged 85 and over live alone in homes they own.
- Divorce has become more socially acceptable in a growing number of countries, resulting in a larger number of single-parent households.
- Improved rights for lesbian and gay people include the right to marry in an increasing number of countries, including Ireland and Brazil. Although some couples have children, more will form two-person households.

Sex ratios

A common feature of age–sex pyramids is an excess of male births over female births. The naturally occurring ratio is 105 boys born for every 100 girls (this imbalance reflects chromosome variations in sperm). In some societies the ratio of male to female children is even more skewed, however. This is usually the result of:

- sex-selective abortion (which is made possible by modern technology allowing couples to discover the gender of a foetus at an early stage of pregnancy)
- female infanticide (the killing of unwanted new-born female babies, also called gendericide, which sadly still happens in some parts of the world).

According to UN statistics, between 100 and 200 million women are 'missing' globally on account of these practices being carried out, both in the past and the present. Table 1.9 shows several countries where the boy:girl ratio of registered births differs significantly from the expected natural balance.

Table 1.9 Countries with a skewed sex ratio showing significantly more males than females

	Average sex ratio at birth 2000–14 (number of males per 100 females)
Albania	112
China	115
Georgia	113
India	112
Liechtenstein	116
Pakistan	109
Taiwan	109
Tunisia	108
Vietnam	111

Source: Population Resource Institute, using UN Census Bureau data

Typically, the favouring of boys reflects the uneven status of men and women in many societies, especially among older age groups. This can be the product of an antiquated legal system that does not allow women to inherit land or wealth (Figure 1.29).

- In the past, there was a very strong cultural preference for boys in China. The country's one-child rule – which was introduced in 1980 (see page 31) – led quickly to the growth of a highly unbalanced population pyramid. In very poor rural areas, rates of female infanticide were sometimes very high. An extreme boy:girl ratio of 4:1 was recorded in some rural counties in the years immediately after the rule was introduced, reflecting the pressing need felt by many poor farming families to gain a male heir. China's most recent census revealed there were 115 males for every 100 females in China as a whole (the actual figure may have been even higher owing to some additional males births going unreported).

- In India's Daman and Diu region, there are 1,000 men for every 618 women according to one report. India's government has long recognized the problem of unbalanced births. In 1994, it passed the Prohibition of Sex Selection Act that prohibits medical professionals from informing expectant mothers of the sex of their foetus. This law is not always upheld, however.

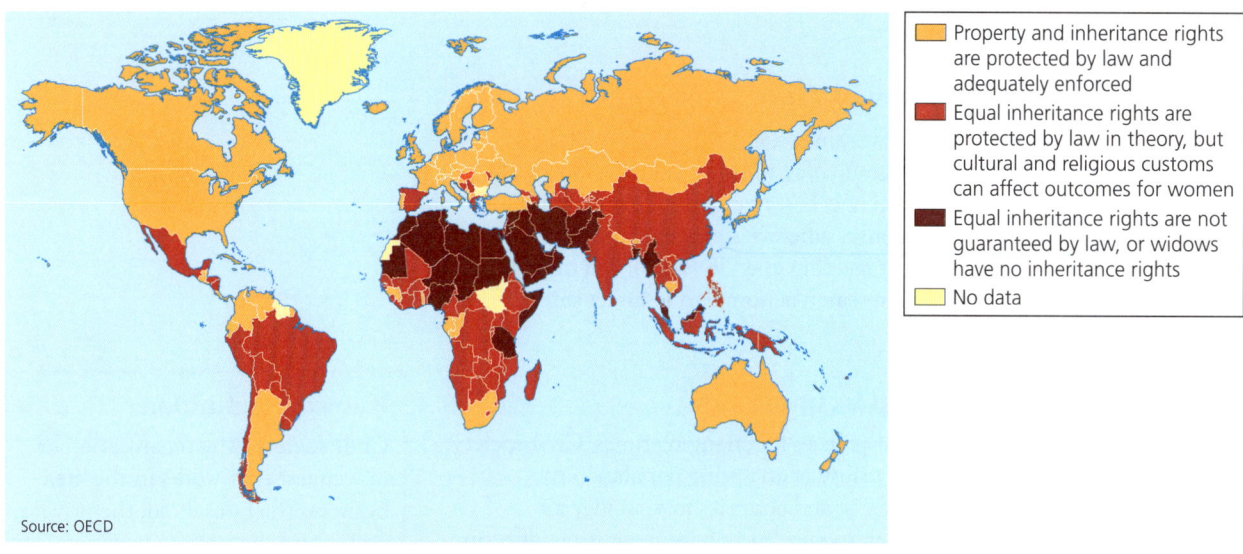

Figure 1.29 The uneven global pattern of inheritance rights for women

Population management policies

The range of population issues affecting societies requires governments to manage them.
- 'Carrot' (reward) and 'stick' (fine or punishment) policies have been used to reduce or increase birth rates in different contexts.
- Economic measures are needed to deal with the challenge of rising elderly dependency; these may be accompanied by other legislation in relation to the retirement age.
- Policies designed to encourage gender equality and improved rights for women may also have the indirect effect of reducing a society's fertility rate.

Policies for ageing societies

Planning for the 'demographic time bomb' is one of the greatest political challenges of our times. Small steps taken by some governments already include:
- minor adjustments to the retirement age
- encouraging workers to invest more of their earnings in personal pension schemes
- stimulating public discussion on whether families should do more to care for older relatives.

However, the measures introduced so far in countries like Germany, the UK and Japan fall far short of what is needed. Governments do not want to raise taxes for younger workers; nor do they want to upset older voters by raising the retirement age too high or too fast. As a result, it remains unclear how many states will be able to fund services for the elderly through taxation by mid-century.

Poland's government has recently acknowledged that its whole state pension system will become unsustainable unless people are prepared to work longer.
- This is a country where pension costs, at 12 per cent of GDP, are already huge. Part of the problem is early retirement, with many people stopping work before they reach 60.
- Another issue is the current shortfall in taxes to pay for pensions: Poland has lost millions of young workers in recent years. Since it joined the European Union in 2004, the total number of Polish emigrants living overseas has risen from 2 to 3 million.

- Although some migrants send home money as remittances, this does not compensate the government for the taxes it might have raised if more of its citizens had remained at home.

When Poland's government warned recently that it might need to raise the retirement age, angry crowds gathered in front of parliament where some set fire to pictures of ministers.

Global interactions between national governments

Increasingly, national governments are working together to tackle the ageing population challenge. The leading G8 group of nations met in December 2013 to discuss the challenge of dementia and how best to coordinate their efforts to tackle it. The global number of dementia sufferers is expected to treble to 135 million by 2050. After their summit meeting, the G8 announced its plans to develop a coordinated international research action plan to target gaps in medical research.

The role of civil society and businesses

It is not just governments that need new policies for changing times. **Civil society** and businesses are adapting to the new reality of an ageing population too.
- Many civil society organizations (CSOs) and charities raise money for vulnerable elderly people or campaign to raise awareness about issues affecting the elderly. Age International is a charity that takes a special interest in helping elderly people in some of the world's poorer countries.
- Swiss TNC Unilever, the world's biggest food and drinks company, is modifying its packaging and products with the elderly in mind: this includes making wrappers easier to open.
- German supermarket chain Kaiser has adapted many of its stores in ways that appeal to older shoppers. This includes brighter lighting, non-slip floors and extra-wide aisles that mobility scooters can navigate easily.
- In Japan, new shopping malls are been designed with the elderly in mind. Older shoppers can access medical clinics and participate in leisure activities targeted specifically at older age groups. Over-60s already account for 40 per cent of Japan's domestic consumption.

Future consumer spending growth in most high-income countries will be driven increasingly by older people. More and more businesses are adapting their products and services accordingly.

Pro-natalist and anti-natalist policies

Some countries have adopted pro-natalist policies that aim to raise the national fertility rate.
- Part of the Polish parliament's response to its ageing population problem has been new legislation that pays women for each child they have in an effort to boost the country's falling population. Under the scheme announced in 2015, every woman receives a monthly payment of 500 zlotys (around US$150) per child (some economists are unconvinced Poland's government will be able to meet the cost of this policy, however).
- Many European countries make monthly child allowance payments to families. In 2015, French mothers received the equivalent of around US$150 per child per month (excluding their first child). Concerned with **underpopulation**, the French government first introduced cash payments for children during the 1930s. In the decades after the Second World War, France had the highest child allowances in Europe, often accounting for around 30 per cent of a family's income.

Other countries have taken the opposite approach by introducing anti-natalist polices designed to curb population growth and prevent **overpopulation** from occurring. The best-known example is China's one-child rule, which lasted from 1980 until 2016. With 1.35 billion people living there in 2015, China is one of the

> **Keyword definition**
>
> **Civil society** Any organization or movement that works in the area between the household, the private sector and the state to negotiate matters of public concern. Civil society includes non-governmental organizations (NGOs), community groups, trade unions, academic institutions and faith-based organizations.
>
> **Underpopulation** A state of imbalance where there are too few people relative to the resources a nation possesses to make effective use of them, lowering quality of life for all as a result.
>
> **Overpopulation** A state of imbalance where there are too many people relative to the resources a nation possesses. The resulting unemployment and insecurity threatens everyone's quality of life.

PPPPSS CONCEPTS

Think about the ways in which the place where you live has been designed with the needs of elderly citizens in mind. What more could be done?

world's two most populated countries (India is the other). Without the one-child rule, it is estimated that China's population might have risen to 1.7 billion, with potentially disastrous consequences.

- In the 1950s, the Chinese fertility rate was high. Under the Communist leadership of Chairman Mao Zedong, women were encouraged to give birth five or six times.
- A devastating famine was linked with the deaths of 25 million people between 1959 and 1961, however. Following this, China's leaders came to believe that the country now had too many mouths to feed; increasingly strict rules began to dictate how many children could be born in China.
- At first, attempts were made to persuade people to have fewer children using the slogan 'Late, sparse, few'. In 1979, persuasion turned to force when the one-child rule was introduced.
- Initially, the rule stipulated that all couples, whatever their circumstances, could have one child only. If people disregarded the policy and had a second child, they were fined US$4,000. This represented a lot of money for poor Chinese families and only very affluent couples could afford to pay. Additional sanctions were used: if a state official had a second child then he or she would lose their job.
- The growth of China's population from 1960 to 2015 is shown in Figure 1.30. The fall in birth rate demonstrates clearly how change was managed by the state.
- After around 2005, the rule was softened progressively. Two babies were permitted for: people living in the countryside; parents whose first child was a girl; parents whose first child had been born with a disability; parents whose only child was killed by the major earthquake in Sichuan Province in 2008.

The gradual relaxation and eventual abolition of the one-child rule in 2016 reflected the fact that its goal had been achieved. In fact, fears of overpopulation have been replaced by the rising challenge of a 'top-heavy' population pyramid with a high dependency ratio (Figure 1.31). This means that more births may be required in the future, rather than fewer.

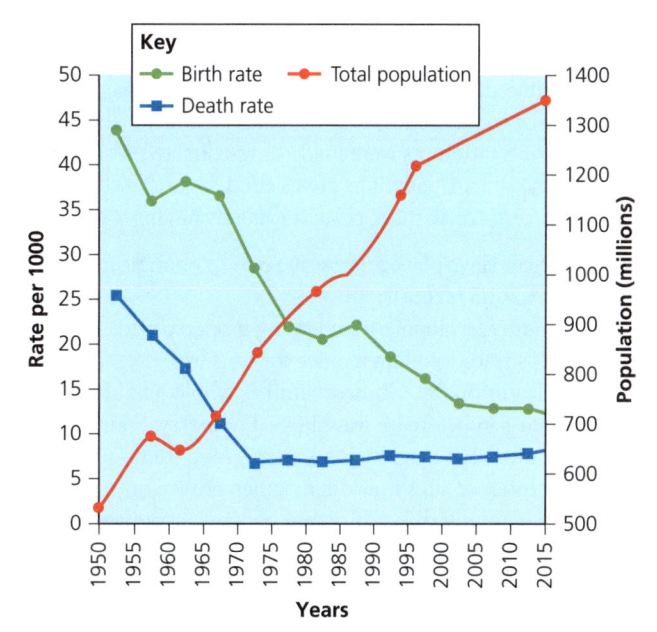

Figure 1.30 Population growth in China, 1950–2015

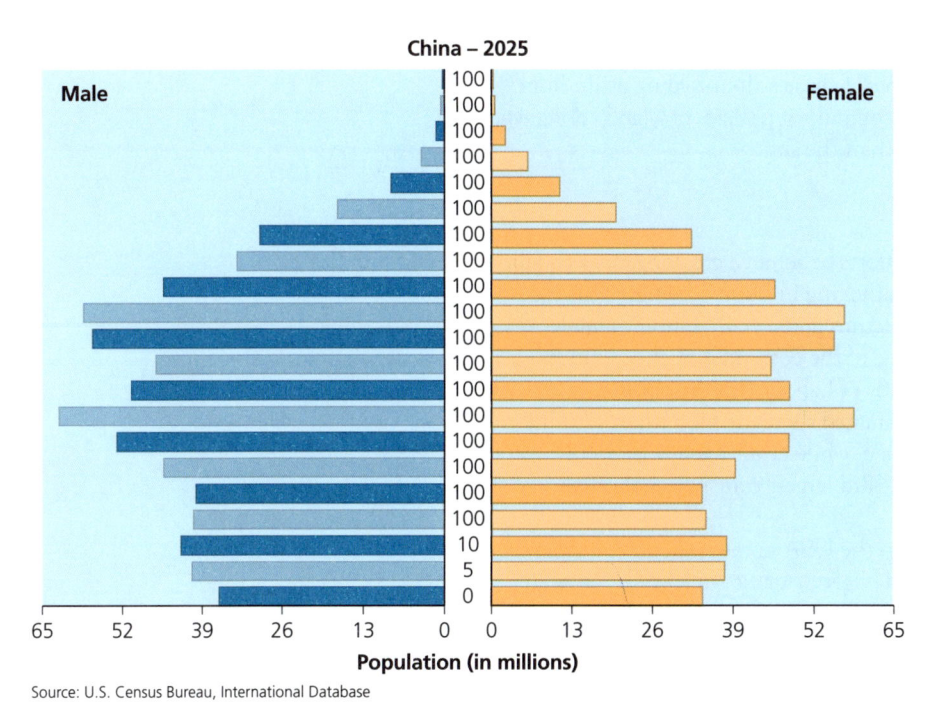

Figure 1.31 China's projected population structure in 2025

Gender equality policies

Prevailing social attitudes about women's role in society play an important part in determining fertility rates and any consequent political attempt to increase or curb fertility. Pro-natal or anti-natal policies can, in turn, have implications for gender equality, which is why they can become controversial. For instance, the Polish government's decision to pay women to have more children has been attacked by gender equality campaigners who say it encourages the stereotyping of women as 'stay-at-home mothers'. The Congress of Women, a women's lobby group, has said the proposal will discourage Polish women from working and strip them of their financial independence.

Changing attitudes in China

Alongside the one-child rule, China's government has tried to change social attitudes (the way people think). Greater gender equality has lowered the fertility rate in many societies by removing the imperative for couples to produce a son who will inherit the family's wealth or name (see page 29). The one-child rule's terrible side effect of female infanticide (see page 28) prompted China's government to fund education campaigns to bring about social change. In some rural districts, attempts were made to re-educate prospective parents using posters bearing the slogan 'daughters are as good as sons' – the hope being that fathers might learn to think more positively about having a girl as their only child.

Other factors have played a positive role in changing social attitudes towards gender in China recently:
- Economic development and rising property prices have helped erode Chinese society's traditional preference for sons by increasing the cost of raising a male heir. Traditionally, Chinese families who could afford to would buy a property for their son before he was allowed to marry. Soaring housing prices in China's major cities have made this an expensive tradition to maintain. No wonder many internet sites now offer women advice on how to conceive girls.
- Accompanying this, China has become a country with strong female role models. Many of the country's very wealthiest individuals are women who run their own businesses, such as Zhang Yin, the head of a recycled paper company, whose personal fortune amounted to US$4 billion in 2015. Eight out of 10 of the world's very richest 'self-made' women are Chinese. Topping the list in 2015 was 'touchscreen Queen' Zhou Qunfei (founder of Lens, a company which makes touchscreens for mobile devices).

Some commentators believe that the roots of these still-unfolding attitudinal changes can be traced back to the early communist policies of Mao Zedong who reportedly believed that 'women hold up half the sky'.

Anti-trafficking policies

One important aspect of international efforts to achieve greater gender equality has been coordinated action to tackle trafficking in human beings. This is a highly gendered form of crime because victims are overwhelmingly female.
- Human trafficking can be seen as 'one of the dark sides of globalization' according to the Institute for the Study of Labour (IZA) in Bonn.
- The US Department of State has estimated there are more than 12 million victims of human trafficking worldwide, while crime agency Interpol categorizes human trafficking as the third largest transnational crime following drug and arms trafficking.
- Female victims of trafficking found in the USA come from 66 countries in different regions (China, Mexico and Nigeria, for example).

The United Nations Convention against Transnational Organized Crime was signed in 2000. The European Parliament has recognized the gender dimension of human trafficking, stating that: 'Data on the prevalence of this crime show that the majority of its victims are women and girls. Sexual exploitation is by far the first purpose of trafficking in women. Most trafficked women are forced into commercial sexual services.' The main EU instrument for fighting trafficking in human beings is Directive 2011/36/EU, adopted in 2011.

The issue of trafficking highlights the continuing need for improved social status for women in parts of the world today.

Population as a resource

The world's seven billionth person was most probably born on 31 October 2011. Global population is projected to reach 8 billion around 2025. Are these landmark events good news or bad news for the global economy and environment? The debate over whether population growth is a blessing or burden – or both – is a very old one.

'People are the ultimate resource,' argued Julian Simon in 1980. Referring to the repeated waste of lives in war during human history, Simon wrote memorably:

> There came to me the memory of reading a eulogy delivered by a Jewish chaplain over the dead on the battlefield at Iwo Jima, saying something like, 'How many who would have been a Mozart or a Michelangelo or an Einstein have we buried here?'

The argument we can infer from this is a compelling one: population growth is beneficial because a larger 'pool' of people is likely to produce more truly exceptional individuals who will innovate and design in unanticipated ways that benefit everyone else enormously. Would you own a smartphone if fewer people had been born in the past and, as a result, iconic inventors such as Bill Gates and Steve Jobs had never lived?

Another advantage of a large population is the sizeable labour force that can help a country to industrialize. People are the **human resources** that allow domestic industries to thrive or act as a magnet for foreign investors. Much of China's 'economic miracle' can be attributed to the enormous size of its cheap labour force.

> **Keyword definitions**
>
> **Human resources** The working-age people found in a place who can generate wealth with the skills and capabilities they possess (dependent on their educational levels, the languages they speak and their capacity to innovate and invent).

The demographic dividend

It is not just population size that matters. Economists and geographers are interested in the benefits brought by a particular type of population structure. A relatively new concept called 'the demographic dividend' (DD) is used to frame the debate. In essence, this is a short-term demographic benefit that is 'cashed in' by a country when it moves from stage 2 to stage 3 of the DTM (see page 14).

- During the early decades of stage 3 of the DTM, a 'bulge' develops in the population pyramid between roughly the ages of 15 and 35. These are the more youthful and energetic cohorts of the working-age population.
- This bulge is a product of the time lag of demographic transition: declining infant mortality (occurring immediately following improvements in health, hygiene and food supply during stage 2) is only followed later by a fall in the number of births (because moral codes governing birth control and sexuality are at first resistant to change).
- The population pyramids for Angola and Vietnam in Figure 1.15 (page 15) show this. In Angola (stage 2), women still give birth to six children and the dependency ratio remains high with many young dependants, whereas Vietnam (stage 3) has recently gained a low ratio of very young dependants compared with working-age groups.

The DD played an important role in the emergence of the 'Asian Tiger' economies during the 1960s and is one reason for China's fast growth in recent decades. In such cases, the movement of large, youthful cohorts into working age, accompanied by falling births, boosts economic and social development because:

- a large, young workforce serves as a powerful magnet for 'footloose' global capital
- workers with fewer children begin making investments, contributing to financial growth
- women become more likely to enter the formal workforce, promoting greater gender equality

- salaried workers quickly become consumers; so global retailers and media corporations view these countries as important emerging markets (MTV and Disney are targeting India in particular).

More recently, numerous emerging economies in Asia, Latin America and, increasingly, parts of Africa have begun to enjoy their own demographic dividend, as Figure 1.32 shows.
- **Asia** Indonesia is currently cashing-in a major demographic dividend (66 per cent of its 240 million people are of working age and 90 per cent are literate).
- **Latin America** Brazil began to enjoy its demographic dividend around 2000; today, it is home to 130 million working-age people out of a total of 192 million. In Colombia, just 33 per cent of the population is dependent; and with a literacy rate of 90 per cent, the future looks promising.
- **Africa** South Africa is beginning to profit from the demographic dividend. Mauritius, Morocco and Algeria recently experienced falling fertility.

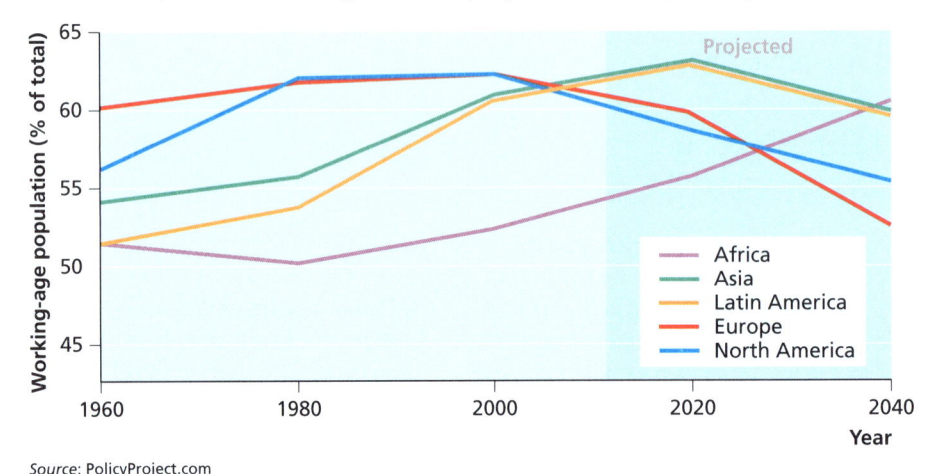

Source: PolicyProject.com

Figure 1.32 Trends in the proportional size of the working-age population by global region, 1960–2040 (projected)

Developed countries, generally, enjoyed the demographic dividend a long time ago, with exceptions. Ireland experienced a late decline in fertility in the 1980s that delivered a demographic dividend in the 1990s (its 'Celtic Tiger' growth phase). The USA, where a relatively large influx of younger immigrants has provided an extended demographic dividend, is an important reminder that migration shapes population structure too.

■ Other factors affecting the demographic dividend

A demographic dividend is *not* always delivered when population structure changes. Some countries fail to make the most of their human resources because of other factors:
- A large working-age population is a wasted opportunity if numeracy and literacy are weak (a concern for some Indian states, where primary and secondary education performance is poor).
- Good governance is essential. Performance of the manufacturing and tertiary sector is determined by economic policies that affect levels of free trade, entrepreneurial activity and inward investment. Providing a framework for growth remains a challenge for some countries (notably Pakistan, according to the British Council).
- In Egypt and Tunisia, where government revenue was, until recently, allegedly siphoned off for the benefit of a small ruling clique, ordinary people received little in the way of 'trickle-down' benefits from their own demographic dividends.
- Political instability jeopardizes the climate for long-term investment. Asian 'success stories' have experienced generally greater political stability (albeit sometimes in non-democratic forms) than parts of Latin America and sub-Saharan Africa. Today, erratic and unpredictable rule can be a powerful deterrent for global capital.
- Youthful out-migration, as in Mexico, may jeopardize receipt of the demographic dividend.

The demographic dividend is, at best, a temporary benefit only. Once several decades have passed, a large number of older workers cease working. With fewer young workers to fund care for the elderly, the dependency ratio begins to rise again, this time involving retirement costs (a 'demographic debt') rather than childcare expense. As we have seen, Japan, which moved through the demographic transition ahead of other Asian nations and benefited from a demographic dividend and economic boom in the 1960s, must now pay for an increasingly elderly population. By 2030, we will know if China, whose recent demographic dividend and 'economic miracle' owe so much to the 1979 one-child rule, can afford to do the same.

CASE STUDY

INDIA'S DEMOGRAPHIC DIVIDEND

India's imminent demographic dividend is one important reason for its new 'superpower' status. Until recently, a dependency ratio close to 0.6 had slowed the country's growth. Now, falling fertility means that by 2030 it could be as low as 0.4. This will mean a higher proportion of population contributing to economic output (and with fewer children to look after, more women entering paid employment). Figure 1.33 compares trends for China and India: as you can see, China has already 'cashed-in' its DD; India is about to.

In theory, the future is bright for India on account of population change:

- While many Asian countries are ageing, approximately half of India's 1.2 billion people are under the age of 26, and by 2020, it is forecast to be the youngest country in the world, with a median age of 29. That means a growing pool of buyers for goods and services, and a growing middle class.
- Between 2006 and 2011, consumer spending in the country almost doubled, from US$500 billion to US$1 trillion. Further growth can be expected.
- India's giant labour force could attract foreign investors who are deterred by rising wages in China (manufacturing labour costs there are nearly four times higher than in India).

However, India's demographic dividend may fail to materialize fully:

- The literacy rate is around 60 per cent (compared with 92 per cent in China). Without improvement, this threatens India's economic development: large numbers of young workers may lack the skills needed to find well-paid employment and contribute to national economy growth.
- India's government has been criticized for its rules and 'red tape', which have sometimes deterred foreign investors. In some industrial sectors, TNCs are obliged legally to work in partnership with a local Indian company if they want to gain market access: not all companies are prepared to work in this way.

As a result, political and social factors could mean that India fails to catch up with China despite its increasingly favourable demographic characteristics.

PPPPSS CONCEPTS

Think critically about the relative importance of demographic, political and social processes of change in determining whether India's economic power may possibly increase in the years ahead.

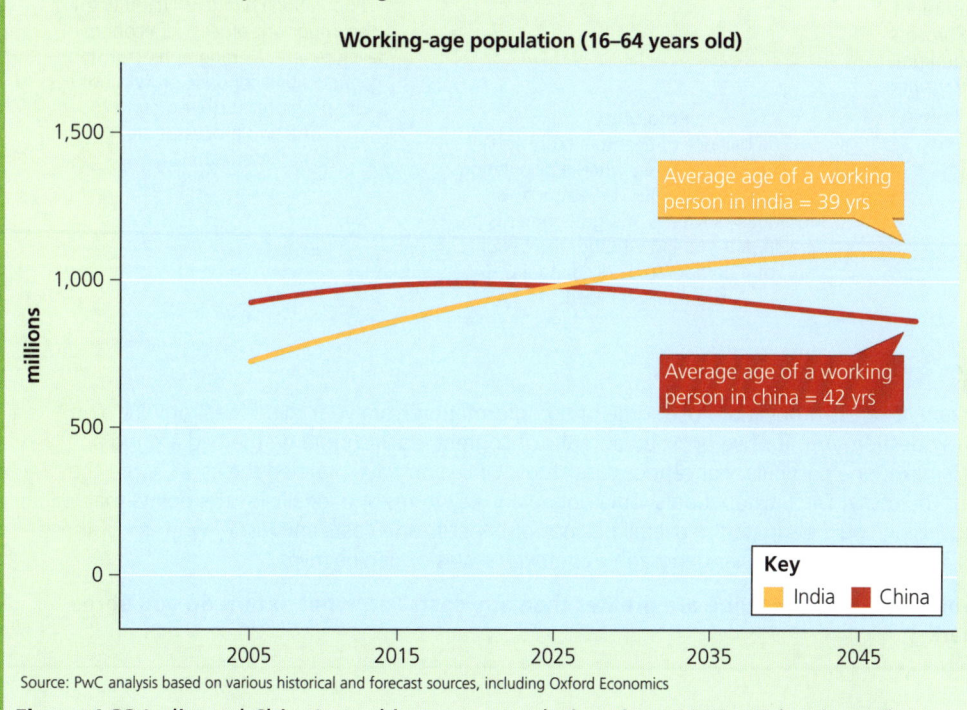

Figure 1.33 India and China's working-age population sizes, 2005–45 (projected)

Unit 1 Changing population

■ KNOWLEDGE CHECKLIST:
- Global and regional ageing trends
- Global and regional trends in family size and sex ratios
- Adaptation policies for ageing societies
- Pro-natalist policies and their effects
- Anti-natalist policies and their effects
- The importance of gender equality policies
- The policy issue of human trafficking
- Human resources and the benefits of population growth
- The demographic dividend (case study of India)

EVALUATION, SYNTHESIS AND SKILLS (ESK) SUMMARY:
- How population change and population policies can help empower women
- How population growth and change can affect the balance of economic power between different countries
- How the issues of population change, gender equality and economic development are interrelated

EXAM FOCUS

MIND-MAPPING USING THE GEOGRAPHY CONCEPTS

The course Geography Concepts were introduced on page vii and have featured throughout the first three chapters of this book.

Use ideas from all of the chapters you have read to add extra detail to the mind map below (based around the Geography Concepts) in order to consolidate your understanding of the topic of changing populations.

Benefits and costs of the demographic transition process

- **Spatial interactions**: Some benefits of demographic transition depend on interactions between different places. China's demographic dividend relied on foreign companies wanting to make use of Chinese labour.
- **Place**: Different places will be experiencing either costs or benefits currently according to their level of development. The costs of an ageing population only begin to be experienced by places when they enter stages 3 and 4 of the DTM.
- **Scale**: Costs and benefits may not be experienced evenly between different local places or countries. While a country as a whole may benefit from demographic transition, some local rural areas may not do.
- **Power**: Demographic changes are linked with changing social attitudes towards gender equality. The process of empowerment for women is integral to any account of demographic transition. The benefits for women of a lower fertility rate are an important part of the discussion.
- **Processes**: Population change is linked with the important process of economic development. Demographic transition is caused by economic growth but it also promotes further growth too. The two processes are interrelated.
- **Possibility**: The balance of eventual costs and benefits is not certain when a population begins to change. The attitudes of individuals, societies and governments are vital in determining whether the benefits of a demographic dividend are eventually realized.

PLANNING AN ESSAY

Below is a sample Section C exam-style essay question. Use some of the information from your mind map, or your own ideas, to produce a plan for this question. Aim for five or six paragraphs of content; each should be themed around a different key point you want to make, or a particular concept or case study. Once you have planned the essay, write the introduction and conclusion for the essay. The introduction should define any key terms and ideally list the points that will be discussed in the essay. The conclusion could evaluate the overall balance of benefits and costs and justify why any of these costs or benefits are especially important for particular geographic contexts, scales or perspectives.

'The benefits of demographic transition over time are greater than any costs.' To what extent do you agree with this statement? [10 marks]

Unit 2: Global climate – vulnerability and resilience

2.1 Causes of global climate change

Revised

Earth's atmosphere has warmed and cooled many times in the past leading to **climate change**. There are two possible reasons why climate change can occur:

- Changing concentrations of **greenhouse gases** (carbon dioxide, methane and nitrous oxide) affect temperature. These greenhouse gases (GHGs) allow sunlight to pass through Earth's atmosphere but trap the heat that is radiated back towards space (short-wave sunlight is absorbed when it hits the ground, or water, and is later re-emitted as long-wave infrared heat radiation). Without GHGs, the Earth would be −18°C (too cold for life to have evolved). If the concentration of GHG rises above its naturally occurring level, however, this is the climatic equivalent of putting extra blankets on a bed. More heat is retained, resulting in a warmer, more energetic and less predictable climate system.
- Various kinds of **external forcing** may occur. This means the amount of sunlight reaching Earth is reduced or increased. This may be because of changes in the Sun's activity or Earth's distance from the Sun, for instance. Any such change in the global energy balance results in a fall or rise in global mean surface temperature (GMST).

> **Keyword definitions**
>
> **Climate change** Any long-term trend or movement in climate detected by a sustained shift in the average value for any climatic element (for example, rainfall, drought, hurricanes).
>
> **Greenhouse gases** Those atmospheric gases that absorb infrared radiation and cause world temperatures to be warmer than they would otherwise be.
>
> **External forcing** A term used to describe processes that impact on Earth's climate system, which originate from outside of the climate system itself, such as variations in solar output.

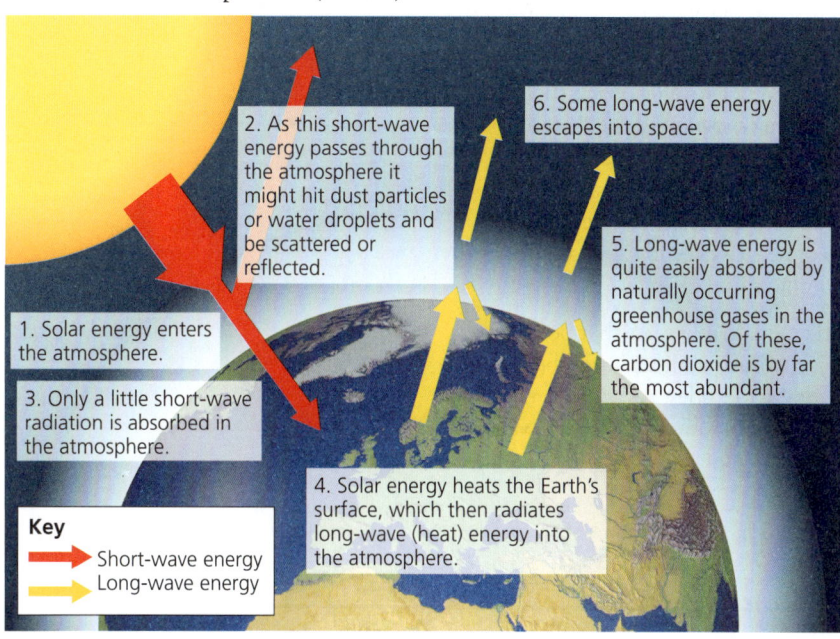

Figure 2.1 The greenhouse effect

Changes in the global energy balance and feedback effects

Revised

A range of evidence suggests that Earth's climate has changed in the past on varying timescales (Table 2.1). Some deviations from the long-term average conditions lasted for millennia; other less dramatic fluctuations have occurred over shorter periods of centuries, decades or years.

Table 2.1 Evidence of naturally occurring climatic variations over time

Evidence	What the data or evidence tell us
Fossil and geological records	Fossils, sedimentary rocks and sediments (containing pollen grains) offer clues about environmental changes over time. The geologically recent appearance and later disappearance of woolly mammoths in what is now Europe indicates that periods of cooling and warming occurred in the past.
Landscape evidence	Evidence for past sea-level changes is provided by raised beaches (landforms which show sea level used to be higher than it is today) and drowned valleys called rias and fjords. From these, we surmise that Earth's ice caps have changed size, along with the volume of water stored in oceans.
Tree rings	Trees produce one new ring of growth per year, during the growing season. A warmer year results in a wider ring. Patterns of slower and faster growth can be observed that allow us to estimate temperatures in the past.
Agricultural records	Historical grape harvest dates have been used to reconstruct summer temperatures in Paris, France, from 1370 to 1879. These contribute to our understanding of the so-called 'Little Ice Age' (a relatively cold phase from around 1300 to 1850) and the 'medieval warm period' (900–1300) that preceded it.

Various theories and data explain why the global energy balance and GMST have fluctuated over time thereby giving rise to environmental changes such as those in Table 2.1. In addition to explaining why world temperatures have sometimes risen or fallen, this chapter also explores how **positive feedback** and **negative feedback** processes help determine the strength and duration of any changes.

- **Positive feedback** loops are knock-on effects in natural systems, which act to *accelerate and amplify* any changes that have already started to occur. When one element of a system changes, it upsets the overall equilibrium, or state of balance, thereby leading to changes in other elements that reinforce what is happening.
- **Negative feedback** occurs when a system adjusts itself in ways that lessen or cancel out the effect of the original change. In this case, feedback has triggered changes in other elements, which act in *the opposite direction* from the initial change. As a result, equilibrium or balance is restored.

Figures 2.2a and 2.2b show examples of positive and negative feedback that help illustrate how these processes operate. The diagrams show two possible ways the climate system could respond to more carbon dioxide being added to the atmosphere (for example, as a result of fossil fuel use). The first scenario suggests accelerated warming as various system changes take place, which strengthen one another. In contrast, the second scenario of negative feedback involves some 'knock-on' effects cancelling out the impact of the initial change.

There is widespread concern among the world's climate scientists that positive feedback effects associated with human emission of carbon dioxide will be far greater than negative feedback effects in the coming decades. As a result, there is a significant risk of a high rise in GMST (of 4–6°C) occurring by 2100 (see page 48).

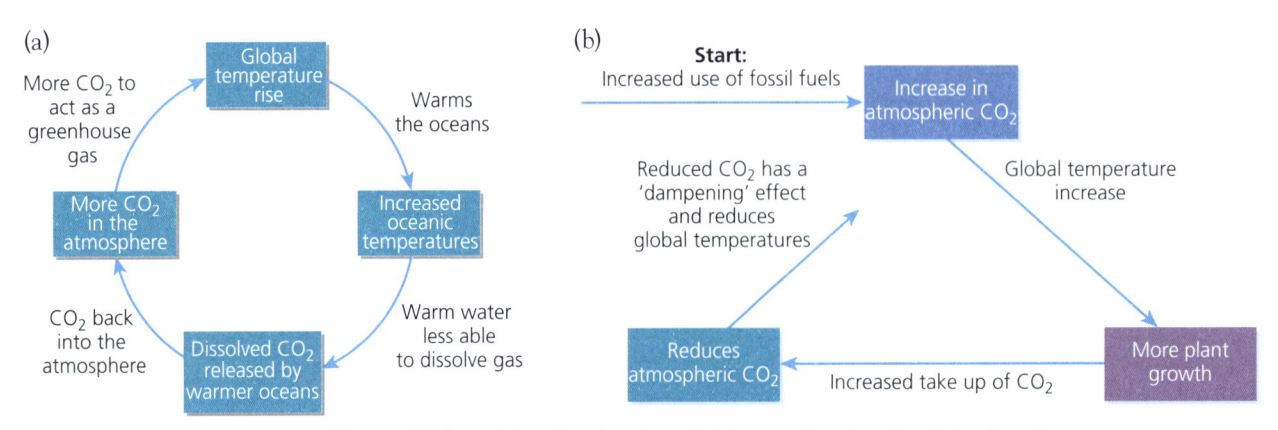

Figure 2.2 The contrasting effects of (a) positive feedback and (b) negative feedback in the climate system

2.1 Causes of global climate change

■ Solar radiation variations

The amount of solar radiation entering Earth's atmosphere can vary, both in the short and longer term. Table 2.2 explains possible reasons for this.

Table 2.2 Reasons for solar radiation variations

Cause	Severity and duration of solar radiation changes
Volcanic emissions and global dimming	Major past volcanic eruptions have led to a short-lived period of global cooling lasting for one or two years only. This is because of ash and dust particles being ejected high into the atmosphere, blanketing Earth and reducing incoming insolation. Krakatoa's eruption in 1873 led to **global dimming** and a fall in GMST of 1°C in 1874.
Changes in solar output	A variety of output cycles have been detected in the amount of energy emitted by the Sun. The most obvious of these is an 11-year sunspot activity cycle. Sunspots are dark areas where intense magnetic storms are happening, which increase solar output (Figure 2.3). A long period with almost no sunspots lasting from 1645 to 1715 is called the Maunder Minimum. It has been linked to the cooler conditions that existed during the Little Ice Age.
Changes in Earth's orbit	Three known cycles occur in Earth's orbit around the Sun, resulting in warming and cooling over long periods of time as incoming levels of solar radiation change (Figure 2.4). Firstly, every 100,000 years, Earth's orbit changes from spherical to elliptical, changing the solar input. Secondly, Earth's axis is tilted at 23.5 degrees, but this changes over a 41,000-year cycle by between 22 and 24.5 degrees, also affecting solar input. Thirdly, Earth's axis wobbles, changing over 22,000 years, bringing further climate change. These orbital cycles are usually termed Milankovitch cycles (the theory of astronomical climate forcing was developed by Milutin Milankovitch).
Cosmic collisions	Large asteroids colliding with Earth may have caused dramatic, though short-lived, climate change in the past. The mass extinction of the dinosaurs is thought to have been caused by a large meteor strike 65 million years ago. Its explosive impact would have thrown up an enormous volume of debris. A vast dust cloud may have blocked sunlight and prevented photosynthesis for up to 10 years. This would have been long enough for the food chains that sustained the dinosaurs to collapse entirely.

PPPPSS CONCEPTS

Think about the importance of the process of feedback for the study of geography. Could population or migration processes be affected by positive or negative feedback, for instance?.

Keyword definition

Global dimming Suspended particulate matter in the atmosphere can reflect solar energy back into space and so have a net cooling effect. This phenomenon can occur naturally because of volcanic emissions but can also be caused by human pollution too – meaning that fossil fuel burning may be both warming and cooling the planet at the same time!

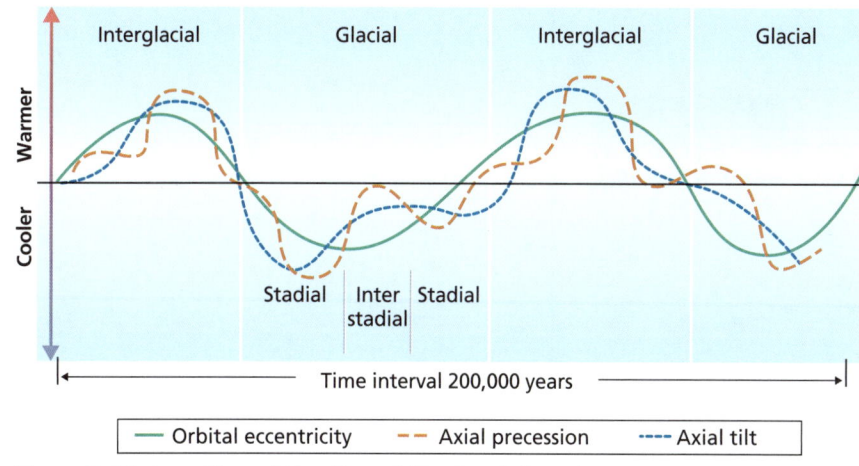

Figure 2.4 Interaction of the three Milankovitch cycles

Figure 2.3 Sunspots

■ Terrestrial albedo changes and feedback loops

During periods when Earth has grown warmer in the past, temperature changes may have accelerated because of the loss of ice cover. Ice has a high **albedo** of around 80 per cent, which means it reflects four-fifths of all incoming solar radiation (Figure 2.5). If some white-coloured sea ice melts in the Arctic, for instance, darker-coloured water will be revealed. The water has a lower albedo

Keyword definition

Albedo How much solar radiation a surface reflects. White surfaces have the highest albedo, or reflectivity.

because of its dark colour and, consequently, it absorbs more heat. As it warms, the water becomes more likely to melt any remaining sea ice. This results in the opening up of more areas of open ocean, and so the process repeats. When the water grows warmer, so too do the air masses that are in contact with it. The result is accelerated warming of the atmosphere.

This positive feedback process means that even a small change in sea ice coverage can have a significant impact on global climate potentially. In theory, even a small reduction in sea ice cover could lead eventually to an ice-free Arctic.

There are periods in Earth's past when the planet has been entirely ice free; at other times the majority of land has been covered with ice. Geologists believe that:
- Earth may have become a giant 'snowball' during two distinct Cryogenian ice ages that occurred around 650 to 750 million years ago
- during much of the Paleocene and early Eocene geological epochs (about 65–35 million years ago), 'hothouse' conditions meant the poles were probably free of ice caps – and crocodiles lived above the Arctic Circle!

Feedback cycles linked with albedo changes occurring both on land and water may have played an important role in producing both outcomes.

Feedback cycles are not always predictable, however. Consider how Figure 2.6 suggests an alternative scenario for the Arctic:

1. Ice melts to expose the darker ocean which absorbs more sunlight.
2. As the ocean warms, there will be increased levels of evaporation.
3. This will create more cloud, especially in the lower atmosphere.
4. The light-coloured cloud has a high albedo and reflects incoming solar radiation (much as the ice used to).
5. As a result, less light reaches and is absorbed by the ocean surface.
6. The temperature of the water falls, as does that of the air mass in contact with it.

This is an example of how negative feedback could help the climate system to self-correct if Arctic ice begins to melt.

Figure 2.5 A white ice surface reflects sunlight

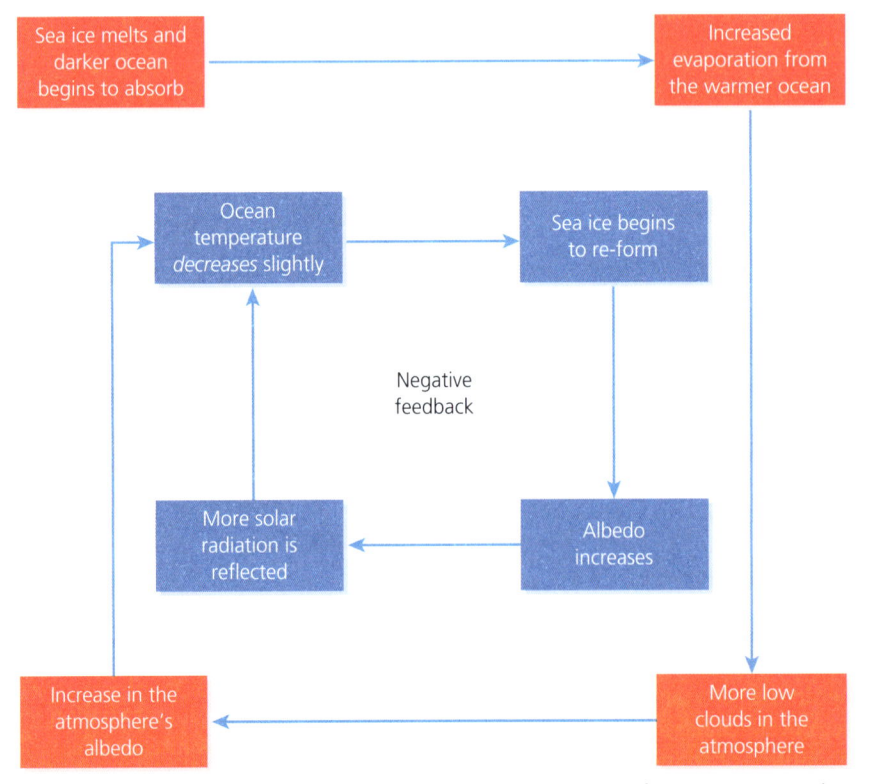

Figure 2.6 How a negative feedback loop operates to nullify environmental change

Methane gas release and feedback loops

Methane is a very powerful greenhouse gas, enormous volumes of which are stored in frozen soils in Earth's **permafrost** regions. Around one quarter of Earth's surface is affected by continuous or sporadic permafrost, including tundra, polar and mountain regions (Figure 2.7). Some researchers estimate that the amount of methane in permafrost equates to more than double the amount of carbon currently in the atmosphere. The effects of permafrost melting may also be magnified potentially over time in positive feedback loops. In theory, this could take Earth's climate beyond a 'tipping point' (the point of no return).

Globally, permafrost covers 23 million square kilometres (mostly in Earth's northern hemisphere). It formed during past cold glacial periods and has persisted through warmer interglacial periods, including the Holocene (the last 10,000 years). However, reports indicate that it is now shrinking at an alarming rate:

- One estimate shows that the tundra climate zone (where most permafrost is found) has shrunk in size by about 20 per cent since 1980.
- According to scientists at NASA, temperatures in Newtok, Alaska, have risen by 4°C since the 1960s, and by as much as 10°C in winter months.
- Based on current evidence, it is likely that widespread melting of permafrost will have occurred by 2100.

> **Keyword definition**
>
> **Permafrost** Ground (soil or rock and included ice) that remains at or below 0°C for at least two consecutive years. The thickness of permafrost varies from less than 1 metre to more than 1.5 kilometres.

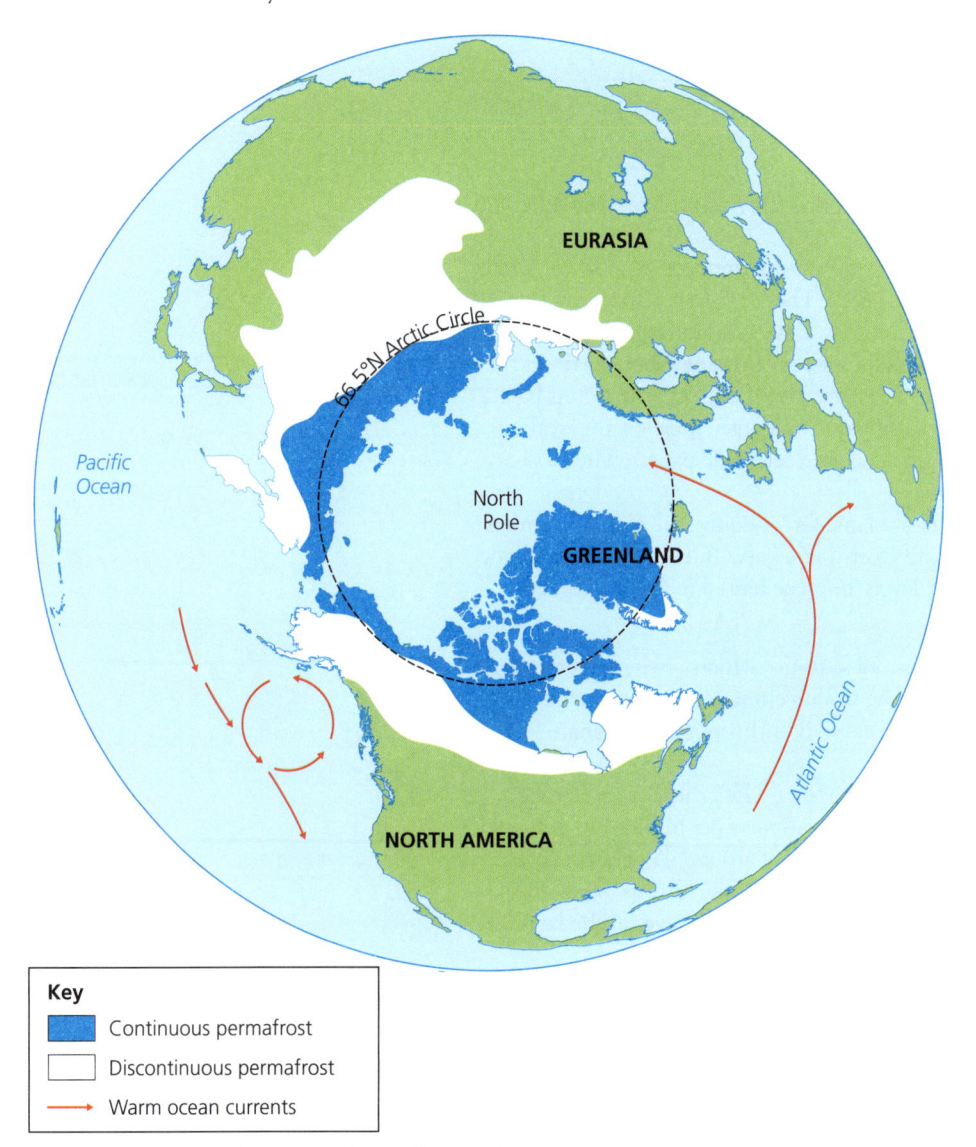

Figure 2.7 Present-day permafrost distribution

The enhanced greenhouse effect

Taken together, a range of evidence suggests that Earth's climate is currently warming and changing, and that humans are most likely to blame for this. In 2015, GMST reached a new record high of +0.87°C relative to the 1951–80 average GMST. The 10 warmest years after 1880 have all occurred since 2000, with the exception of 1998 (Figure 2.8).

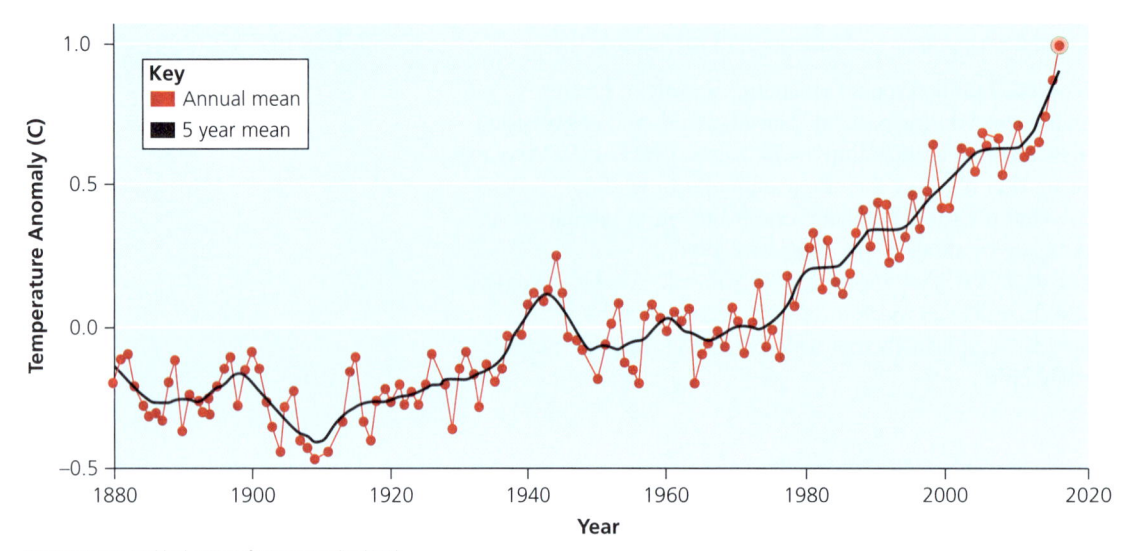

Source: NASA's Goddard Institute for Space Studies (GISS)

Figure 2.8 Global temperature anomaly trends (deviation from the 1951–80 average)

The US National Oceanic and Atmospheric Administration (NOAA) reported 'unmistakable' signs of global warming in 2010. The UK's Met Office says that the signs of warming are now so clear that they 'have human fingerprints' all over them. In 2013, the Intergovernmental Panel on Climate Change (IPCC) published its fifth assessment report in 23 years. The IPCC organization is based in Geneva. Its reports, many thousands of pages in length, draw on the work of over 800 scientists and hundreds of peer-reviewed scientific papers. The 2013 assessment stated that:

- It is 'virtually certain' humans are to blame for 'unequivocal' global warming.
- Atmospheric concentrations of the greenhouse gases (GHGs) carbon dioxide, methane and nitrous oxide are at levels 'unprecedented in at least the last 800,000 years'.

The rate of change is exceptional in historical terms, which is why the prevailing viewpoint held by the large majority of the world's climate scientists is that humans are to blame. In other words, the cause is anthropogenic (human). The key steps in the argument are as follows:

- Carbon dioxide emissions have been rising since 1750, the start of Europe's Industrial Revolution, from a level of 280 ppm (parts per million) to 406 ppm at the start of 2017. This represents an increase of around 45 per cent. Even more worryingly, if you convert other GHGs (methane, nitrous oxide) into their equivalent amounts of CO_2, then you find that we have reached a level in excess of 470 ppm of CO_2 equivalents).
- This is because population growth and economic development have led to: worldwide use of carbon-rich fossil fuels as an energy source; widespread deforestation; and enhanced methane emissions derived from livestock (bovine flatulence) and the decomposition of organic wastes in landfill sites.

- A steep trend line can be viewed in carbon dioxide data: there was gradual growth up to the early 1900s and then very rapid growth. Recent measurements have been taken offshore at Mauna Loa every year since the 1950s and show a rise each year of around 1–2 ppm. Older evidence comes from 'fossil air' trapped in ice. Permanent ice in high mountains and polar ice caps has built up from snow falling over hundreds of thousands of years. Cores 3–4 kilometres long have been extracted from the Vostok ice sheet in the Antarctic, producing ice that is more than half a million years old and containing bubbles of ancient air. After analysing the evidence, scientists believe that carbon dioxide and methane concentrations are now higher than at any time in the last 800,000 years (Figure 2.9).

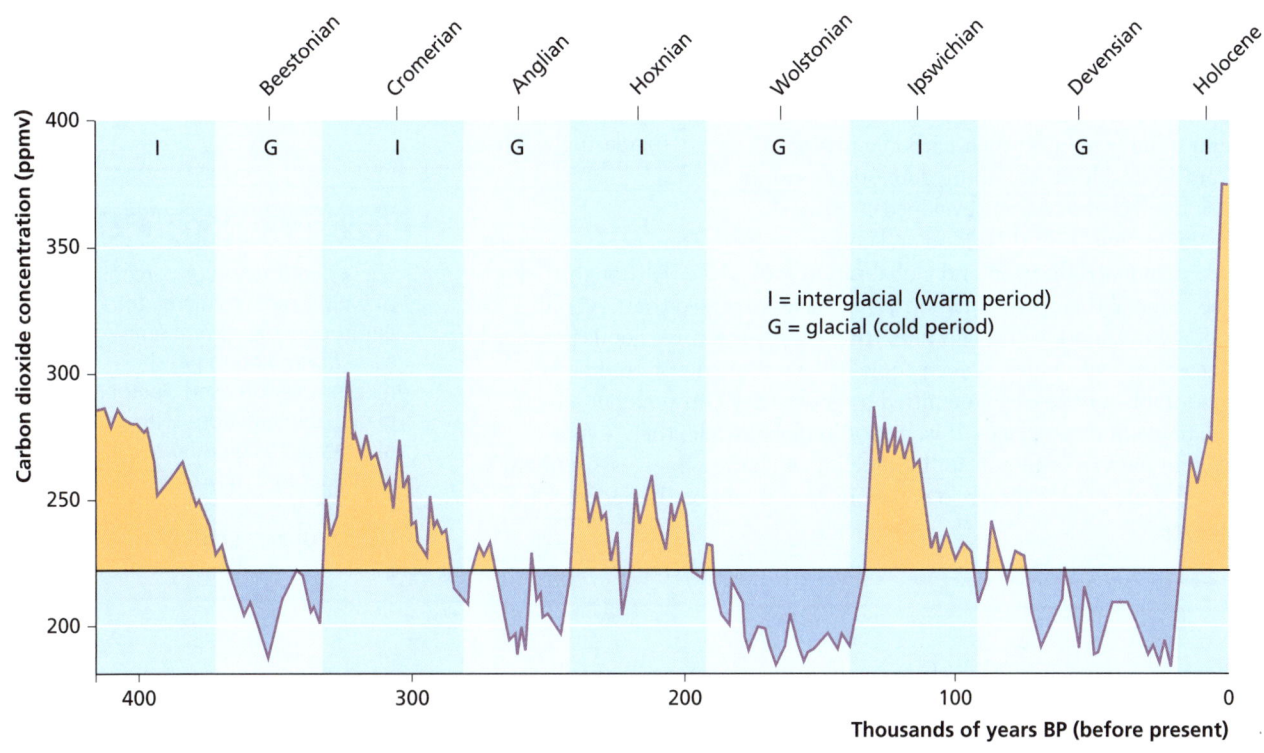

Figure 2.9 Atmospheric carbon dioxide concentrations measured from (a) Vostok ice cores and (b) recent Mauna Loa data

Global patterns of carbon emissions

You will probably be familiar already with the general pattern of global carbon emissions. The highest polluters in terms of total carbon output are high- and middle-income countries with large populations, including China, the USA, India, Russia, Japan and Germany (Figure 2.10).

When comparing countries there are three important things to remember:
- Data showing **per capita carbon footprints** reveal a very different pattern. The highest figures belong to oil-rich Middle Eastern states including Qatar and Kuwait. The USA and EU countries have relatively high per capita carbon footprints but so too do many Caribbean countries where wood is burned widely. China has a relatively low per capita carbon footprint.

Keyword definition

Per capita carbon footprint
The amount of carbon dioxide emissions an average person in a country is responsible for as they go about their everyday life.

- Current data show the present-day **anthropogenic carbon flow** generated by different countries. However, data showing which countries are responsible for most of the **anthropogenic carbon stock** already in the atmosphere reveal a very different pattern; Germany, for example, may be to blame for around 6.9 per cent of all anthropogenic carbon emissions dating back to 1750 (according to the World Resources Institute). In 2015, Germany accounted for only 2.3 per cent of current carbon flow. Therefore, our perception of which countries should be held most to account for anthropogenic climate change may alter depending on whether we look at carbon stock or carbon flow data – or both.
- We need to think carefully about globalization and trade too. Some high-income countries claim to have reduced their carbon emissions in recent years, However, falling domestic emissions mask the fact that many developed countries now import much of their food and consumer goods from other countries since **deindustrialization** and globalization took place. Although they have reduced their own emissions, they have increased their overall carbon consumption through importing energy-intensive goods produced in emerging economies. Some people believe countries should be held accountable for the carbon emitted by goods they consume but which were made in other places. If we factor in global trade, then a very different pattern of responsibility for climate change emerges. Germany's annual share of global carbon emissions rises to almost 10 per cent, for instance.

> **Keyword definitions**
>
> **Anthropogenic carbon flow** The current amount of carbon emission released annually by a country (for example, due to fossil fuel burning and cement making) produced in each nation. The figure can be adjusted upwards to factor in the carbon equivalents of other greenhouse gas emissions (methane and nitrous oxide).
>
> **Anthropogenic carbon stock** The total size of the store of anthropogenic carbon emissions released into the atmosphere since industrialization began around 1750.
>
> **Deindustrialization** The loss of traditional manufacturing industries in some high-income countries due to their closure or relocation elsewhere. Since the 1960s, many industries have all but vanished from Europe and North America. Instead, they thrive in Asia, South America and, increasingly, Africa.

> **PPPPSS CONCEPTS**
>
> Think about how our perceptions of who is most to blame for anthropogenic carbon emissions varies according to which measurement is used. Factoring in global interactions with other places where commodities are manufactured may be the best way to reveal the true carbon footprint of high-income countries.

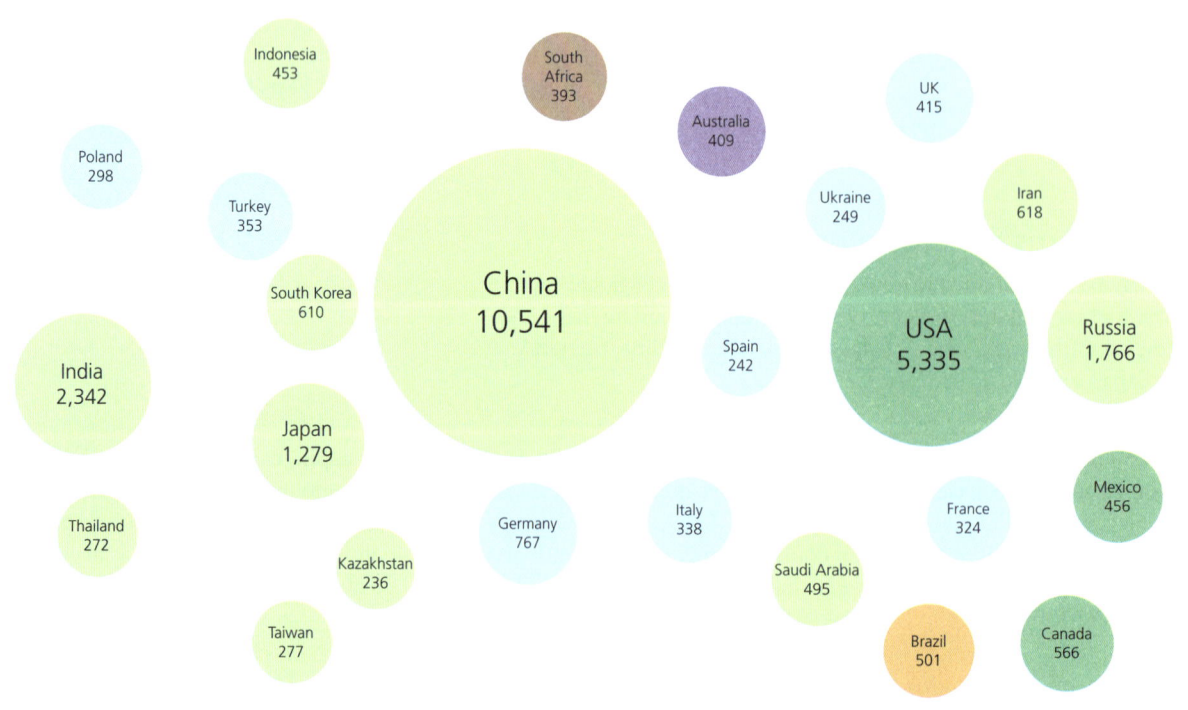

Figure 2.10 Global carbon emissions for selected countries, 2014 (million metric tonnes)

Table 2.3 analyses the role and actions of selected individual states and groups of countries in relation to global carbon and GHG emissions.

Table 2.3 The role and actions of states and groups of countries in relation to global emissions

	Role and actions
USA	■ The USA is the world's second biggest emitter, generating just under one fifth of all anthropogenic GHGs. This is despite its population size being less than 5 per cent of the global total. ■ In recent years, its carbon emissions have fallen by around 10 per cent. The main reason for this is a shift from coal to gas burning within the USA's energy mix. A gas-fired plant produces half the emissions of a coal-fired one. ■ On a per capita basis, the USA still has a carbon footprint five times higher than China. Under the Obama administration, the USA began taking greater action to address its emissions. However, the Trump administration is more skeptical of the need for action.
China	■ China became the world's largest carbon emitter in 2007 on account of widespread industrialization since the 1970s (see page 83). A massive programme of poverty alleviation (requiring more, not less, energy) remains the country's priority. ■ China now contributes around one quarter of world emissions while accounting for one fifth of world population. Its emissions rose by almost 10 per cent in 2011 alone, primarily due to higher coal consumption. ■ China's leaders want to reduce the rate at which their emissions rise and have made a (non-binding) pledge to reduce the **carbon intensity** of the country's growth by adopting more **renewable energy** into their mix. For instance, China spent US$10 billion on wind turbines in 2010 – about half of all global spending. Such actions will not lead to a cut in total emissions but will curb the rate at which China's emissions grow.
India	■ India is home to nearly one fifth of the world's population. Currently, it contributes more than 5 per cent of global CO_2 emissions.
Qatar	■ The oil-rich desert state of Qatar has exceedingly high per capita emissions. The country's great oil wealth is used to fund high energy usage, including lavish use of air conditioning.
Japan and Germany	■ Japan's emissions have increased by several per cent recently as a result of a substantial increase in the use of fossil fuels in power generation. This is part of a reaction against the use of nuclear power following the accident at Fukushima in 2011. Germany is also phasing out nuclear power but hopes to restrict emission rises by adopting renewable energy sources.
Low-income countries	■ Many of the world's least developed countries (LDCs), such as Somalia, continue to make a negligible contribution to anthropogenic GHG emissions, although economic changes in some African countries, such as Nigeria and Kenya, mean that energy consumption there is rising. ■ People in some poor countries may have a high per capita carbon footprint because of their reliance on wood-burning stoves as a source of heat and energy for cooking.

Keyword definitions
Carbon intensity The amount of CO_2 emitted per unit of GDP. If a country's carbon emissions rise less slowly than its GDP is increasing, this suggests some action is being taken to reduce emissions at the same time as industrial output is increasing.

Renewable energy Wind, solar and tidal power sources that result from a flow of energy from the Sun.

■ The complexity of the dynamic climate system

This chapter has demonstrated that climate change science is far from simple. There is uncertainty over the operation of feedback loops. Nor can we predict with any certainty what kinds of economic and demographic changes will take place globally or in particular countries, as Unit 1 showed. As a result, climate change presents us with a **wicked problem** which – because of its complexity – defies attempts to establish exactly what its effects would be. Unfortunately, this uncertainty is seized upon by climate change skeptics as a reason to avoid taking any action to reduce GHG emissions.

Other factors introduce greater complexity yet into climate change models. Natural climatic cycles like the **El Niño Southern Oscillation (ENSO)** and the North Atlantic Oscillation (NAO) are independent phenomena that operate naturally on shorter, decade-long timescales. They introduce year-to-year climatic variability into numerous different local contexts. Both phenomena can bring colder conditions temporarily to places that are predicted to get warmer on account of long-term climate change. This complicates our understanding of long-term climate change and can undermine public faith in IPCC expert predictions. People who lack deeper scientific understanding of the issues may – quite understandably but erroneously – view one unusually cold winter as 'irrefutable proof' that the planet cannot be warming up.

Figure 2.11 provides a useful summary of some of the many different influences on climate change processes and the high levels of complexity and uncertainty that affect future predictions of climatic change.

> **Keyword definitions**
>
> **Wicked problem** A challenge that cannot be dealt with easily owing to its scale and/or complexity. Wicked problems arise from the interactions of many different places, people, things, ideas and perspectives within complex and interconnected systems.
>
> **El Niño Southern Oscillation (ENSO)** A sustained sea surface temperature anomaly across the central tropical Pacific Ocean. It brings a change in weather conditions that can last from two to seven years. Along with La Niña events, El Niño events are part of a short-term climate cycle that brings variations in climate but only for a few years.

> **PPPPSS CONCEPTS**
>
> Think about the complexity of the climate system. Why might different people living in different places have varying perspectives on whether or not climate change is a 'fact'?

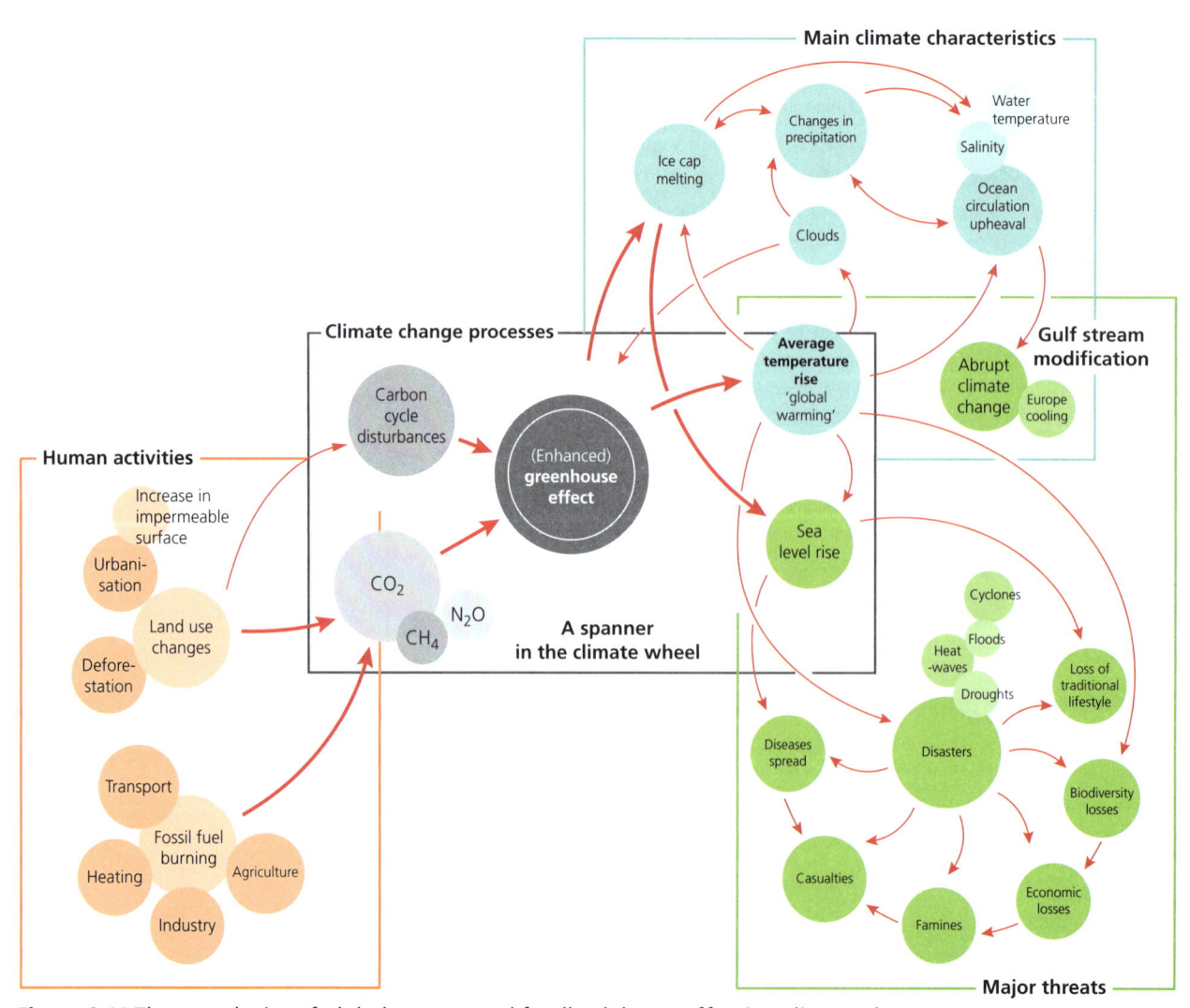

Figure 2.11 The complexity of global systems and feedback loops affecting climate change
Source: Cameron Dunn

2.1 Causes of global climate change

■ KNOWLEDGE CHECKLIST:
- The natural greenhouse effect and the energy balance between incoming short-wave radiation and outgoing long-wave radiation
- Evidence for changes in the global energy balance over time
- Causes of solar radiation variations over time, including sunspots, orbital changes and global dimming due to volcanic eruptions
- The complexity of positive and negative feedback loops, and their effects on global climate
- Terrestrial albedo changes and associated feedback loops
- Methane gas release from permafrost and associated feedback loops
- Human activity and the enhanced greenhouse effect
- Global patterns and trends for anthropogenic greenhouse gas emissions

EVALUATION, SYNTHESIS AND SKILLS (ESK) SUMMARY:
- How study of the climate system and feedback loops reveals the complexity of processes of change
- How global interactions make it hard to determine which places have the greatest responsibility for anthropogenic greenhouse gas emissions

EXAM FOCUS

SUPPORT OR CHALLENGE?

Below is a sample Section C exam-style essay question. It asks how far you agree with a specific statement. Below this is a series of general statements that are relevant to the question. Using your own knowledge and the information in this chapter (and possibly some material from earlier chapters relating to population growth and economic development), decide whether these statements support or challenge the statement. In each case, tick the appropriate box. Then make notes detailing why you have made these choices.

'The causes of climate change are mainly natural in origin.' To what extent do you agree with this statement? [10 marks]

STATEMENT	SUPPORT	CHALLENGE
Changes in Earth's orbit have led to long-term changes in climate, including ice ages.		
The rise in carbon dioxide since 1750 is unprecedented.		
Volcanic activity may lead to a temporary period of global dimming.		
Industrialization, deforestation and cattle farming are most likely responsible for recent temperature rises.		
Sunspot activity can influence climate for decades or longer.		
Positive feedback naturally amplifies climate change irrespective of whether human factors are to blame.		

SPECTRUM OF IMPORTANCE

Below is another sample Section C exam-style essay question (also on the topic of climate change), and a list of general points that could be used to answer the question. Use your own knowledge and the information in the chapters you have read so far to reach a judgement about the importance of these general points to the question posed. Arrange them in order of their relative importance along a spectrum. Having done this, write a brief justification of your placements, explaining why some of these factors are more important than others. The resulting diagram could form the basis of an essay plan.

'Population growth is the main reason why climate change is unavoidable in the 21st century.' To what extent do you agree with this statement? [10 marks]

1. The industrialization of Europe and North America in the 1800s
2. The industrialization of China and India since the 1970s
3. The growth of the global middle class
4. Positive feedback effects leading to permafrost melting
5. Positive feedback effects leading to Arctic ice melting
6. The failure of world leaders to take action to stop climate change
7. The need for low-income countries to catch up with high-income countries
8. The predicted increase in world population from 7 billion to 9 or 10 billion
9. Widespread forest removal in countries like Brazil and Indonesia

2.2 Consequences of global climate change

Revised

Climate change brings a range of new or amplified risks for the environment and vulnerable places and people. Predicted impacts on the physical environment can be modelled in a structured way by examining in turn possible consequences for:

- the **hydrosphere** (the world's water cycle and stores). Major ice sheets have lost mass, land-based glaciers have shrunk and Arctic sea ice cover has fallen significantly since 1979.
- the **atmosphere** (and the size of its carbon store). According to the IPCC, by 2100, global temperature is 'likely' to have risen by more than 1.5°C relative to 1850 and may rise by more than 4°C (Figure 2.12). The IPCC is 'virtually certain' of further permafrost melting, which will release further methane into the atmosphere.
- the **biosphere** (the world's ecosystems, flora and fauna). The IPCC is 'virtually certain' that the upper 700 metres of the ocean have warmed since 1971. In 2016, sea temperatures in the northern section of the Australia's Great Barrier Reef rose 2–3°C above the normal peak of about 30°C, due to the strong El Niño weather system and a continuing trend towards global warming. A report by James Cook University showed that two-thirds of corals in one part of the reef have died as a result of coral bleaching in overly warmer water.

Interactions between the hydrosphere, atmosphere and biosphere are complex and hard to predict. At the global level, the atmosphere, hydrosphere and biosphere can be considered to be an open system that forms part of a chain. Interlocking relationships among these three subsystems (such as the ocean's ability to absorb carbon dioxide) are what make climate change complex and difficult to model, as Chapter 2.1 showed. This section explores interactions that scientists believe are likely possibilities.

Remember that even if all human GHG emissions (new carbon flows) were to be halted tomorrow, some degree of change remains inevitable. This is because up to 40 per cent of the anthropogenic carbon stock already in the atmosphere will remain there for more than 1,000 years. It is therefore 'virtually certain' that some level of warming will continue well beyond 2100.

PPPPSS CONCEPTS

Think about the different possibilities shown in Figure 2.12. What factors might affect whether the lowest or highest scenario turns out to be correct? Why is a range of projected temperature changes shown in each case?

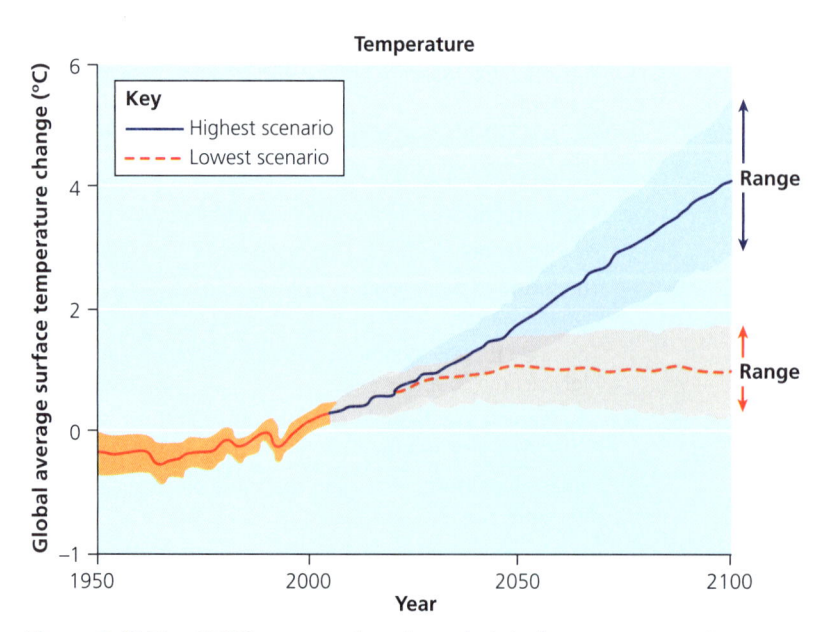

Figure 2.12 The IPCC's range of projected global temperature changes for 2100

2.2 Consequences of global climate change

Water and carbon storage

Approximately 97 per cent of all water on Earth is oceanic and saline. Most of the fresh water that makes up the remaining 3 per cent is locked up in land ice, glaciers and permafrost. This makes up the majority of the Earth's **cryosphere** (Figure 2.13). Aside from cryospheric water, a relatively tiny amount of fresh water is stored in the form of groundwater, lakes, soil, wetlands, rivers, biomass and the atmosphere.

> **Keyword definitions**
>
> **Cryosphere** Those portions of Earth's surface where water is in solid form.
>
> **Mass balance** The difference between the amount of snowfall gained by a glacier or ice sheet, and the amount of ice lost through the processes of calving (blocks breaking off) and/or melting.

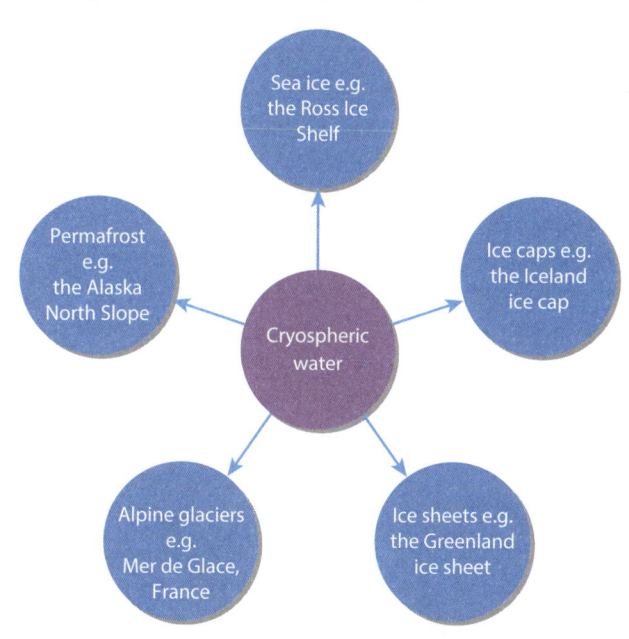

Figure 2.13 The five locations of water in the cryosphere

■ Water storage in the cryosphere

The available evidence suggests that the cryosphere is, in general, becoming reduced in size. Patterns and trends are not uniform, however. In Antarctica at altitudes above 400 metres, glaciers are not thinning and may even be gaining mass. This is because snowfall is increasing across the Antarctic Peninsula (because warmer air can hold more moisture). However, this increased snowfall in upper regions is not enough to offset melting at lower attitudes. Overall, across the whole Antarctic Peninsula, around four-fifths of all glaciers are receding.

Table 2.4 examines changes affecting different frozen water stores within the cryosphere. In many cases, there is a reported reduction in the **mass balance** of the ice store. The human implications of the loss of some or all of these stores include sea-level rise (page 51) and the threatened loss of water supplies in some heavily populated world regions (page 95).

Table 2.4 Evidence of changes in cryosphere storage

Store	Changing characteristics
Ice caps and Alpine glaciers	■ Ice caps occur all over the world, from the polar regions to mountainous areas such as the Himalayas, the Rockies, the Andes and the Southern Alps of New Zealand. The Furtwangler Glacier on Kilimanjaro is Africa's only remaining ice cap. It is melting rapidly and may soon disappear. ■ Alpine glaciers are thick masses of ice found in deep valleys or in upland hollows. These glaciers are particularly important in the Himalayas where meltwater from 15,000 Himalayan glaciers supports perennial rivers such as the Indus, Ganges and Brahmaputra. In turn, these are relied on by hundreds of millions of people in South Asian countries (Pakistan, Nepal, Bhutan, India and Bangladesh). ■ Glacial retreat is clearly evident in aerial photographs of retreating glacier snouts and vanishing ice, for instance at Wolverine Glacier in Alaska or Argentina's Upsala Glacier (Figure 2.14). China's Urumqi No. 1 glacier has reportedly lost more than 20 per cent of its volume since 1962.
Permafrost	■ The vast permafrost ring around the Arctic Ocean has already begun to thaw in places where temperatures have risen by several degrees since the 1960s (see page 57). ■ This melting is releasing large amounts of carbon dioxide and methane, potentially affecting the global climate further.

Store	Changing characteristics
Greenland and Antarctic ice sheets	■ An ice sheet is a mass of glacial land ice extending more than 50,000 square kilometres. The two major ice sheets on Earth today cover most of Greenland and Antarctica. Together, they contain more than 99 per cent of the freshwater ice on Earth (the Antarctic ice sheet covers the same area as the USA and Mexico combined; the Greenland ice sheet is much smaller). ■ Overall, Antarctica is losing more ice than it gains each year. The deficit is around 70 gigatonnes per year (1 gigatonne is 1 billion tonnes). Warmer surface air temperatures around the northern Antarctic Peninsula are having a dramatic effect. ■ If the Greenland ice sheet melted altogether, scientists estimate that sea level would rise about 6 metres. If the Antarctic ice sheet melted, sea level would rise by about 60 metres.
Greenland and Antarctic ice shelves	■ Ice shelves are floating platforms of ice that form where ice sheets and glaciers move out into the oceans; they exist mostly in Antarctica and Greenland, as well as in the Arctic near Canada and Alaska. ■ The Antarctic ice sheet is melting from below, as increased upwelling of relatively warm, deep water comes into contact with the underside of the ice shelves.
Arctic sea ice	■ Within the Arctic Circle lie northern parts of Russia, Canada, Alaska (USA), Greenland and Scandinavia, as well as the Arctic Ocean, parts of which remain covered with sea ice all year. ■ Future temperature rises are projected to be greatest at high latitudes and in recent decades the Arctic has heated twice as fast as the rest of the world. ■ Some experts predict the region will be entirely ice free by 2050. Temperatures 20 °C higher than usual in late 2016 suggest it may occur even sooner.

Figure 2.14 Retreat of Argentina's Upsala Glacier, 2002–13

■ Carbon storage in oceans, ice and the biosphere

Carbon moves naturally from one store (such as vegetation biomass) to another (the atmosphere) in a continuous cycle. The processes by which the carbon moves between these stores are known as transfers or fluxes. The recent increase in GMST has begun to affect the pattern of global carbon stores and fluxes.

Given that water covers two-thirds of Earth's surface, the ocean plays a very important part in the carbon cycle.
- According to data collated by the Global Ocean Data Analysis project (GLODAP), the oceans contain around 40,000 gigatonnes of carbon in the form of dissolved carbon dioxide, marine organisms and dissolved organic matter.
- It is estimated that about 30 per cent of the CO_2 that has been released into the atmosphere has diffused into the ocean through direct chemical exchange. Dissolving carbon dioxide in the ocean creates carbonic acid.
- If this had not occurred then atmospheric concentrations of carbon dioxide would be even higher than they are.
- Unfortunately, the increased acidity of ocean water has harmful effects for coral reef organisms (see page 59) and other sensitive species.

In addition, the pattern of carbon stored in ice and ecosystems is changing:
- Permafrost ice melting releases methane gas (CH_4), which is one part carbon and four parts hydrogen.
- The total amount of carbon stored in the terrestrial (land) biosphere has been estimated at just over 3,000 gigatonnes of carbon. The largest amount is held in tropical and temperate forests. In some places, canopies of tall broad-leaved trees – and their associated under-canopies and shrub layers – have a growing risk exposure to wild fires and so-called 'megafires' that are reportedly increasing in frequency (Figure 2.15). According to one study based on NASA satellite data and IPCC climate projections, there will most likely be a 35 per cent increase in the days with high danger of fire across the world by 2050. Vegetation in eastern states of the USA, southeastern Australia, the Mediterranean and southern Africa is at greatest risk.

Figure 2.15 A predicted increase in fires and megafires could reduce the amount of carbon stored in biomass

■ Sea level rise

Historically, sea levels have always been higher during warm periods and lower during cold phases. During the most recent Pleistocene Ice Age, the UK and France were joined together following a **eustatic** fall in sea level of around 100 metres (the English Channel was drained entirely of water).

A warmer climate results in deeper oceans for two reasons. Both are related to temperature increases, but in different ways:
- **melting** of land-based glaciers and ice caps (and so increased runoff of water from the land to the sea)

> **Keyword definition**
> **Eustatic** A worldwide change in average sea level resulting from a warming or cooling climate affecting the volume and/or depth of water in the oceans.

Unit 2 Global climate – vulnerability and resilience

- **thermal expansion** of the oceans (water actually expands slightly in volume as its temperature rises, like liquid in a thermometer: in a warmer world the sea level would rise even if no ice had melted).

Together, thermal expansion of the oceans and worldwide glacial melting are projected to bring a series of eustatic sea-level rises (Figure 2.16).
- Already these combined processes are giving rise to a total global average sea-level rise of approximately 3 millimetres per year.
- In total, global sea level has risen by 200 millimetres since 1900.
- The IPCC projects a world sea-level rise of 260–820 millimetres by 2100, mostly due to thermal expansion.
- Significant glacier and permafrost meltwater runoff will produce another metre of sea-level rise by around 2200 (even if GHG emissions are cut significantly).
- Without any action to cut emissions, large-scale melting of the Greenland and Antarctica ice sheets could contribute 7-metre and 60-metre rises respectively (but this would take hundreds of years).

It has long been recognized that any sea-level rise will have a bigger effect on some countries than others. The worst impacts will be for places that are coastal and low-lying, such as the Maldives islands (Figure 2.17). Other vulnerable places include estuaries or islands exposed to storm surges driven by cyclones, such as the Thames estuary and the Philippines. The places affected worst may be those suffering additionally from local sinking of the land. In these regions, the net effect will be an even greater rise in the level of water.
- The Ganges delta, in Bangladesh, is sinking by 10 millimetres a year as sediments settle. Some of the world's poorest and most vulnerable people live here.
- Southern England is sinking by several millimetres a year due to a geological process called isostatic tilting, triggered by the end of the last Ice Age.

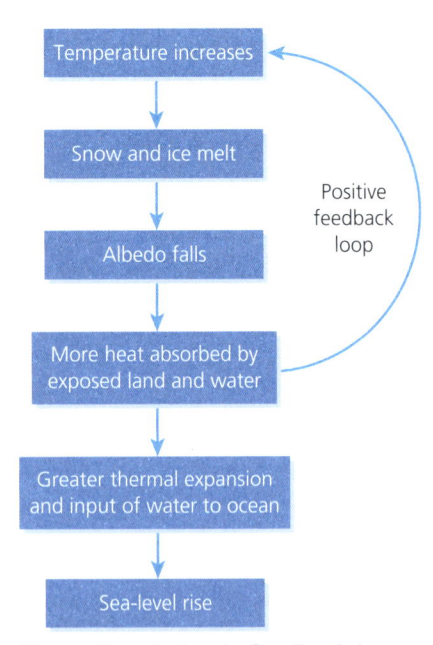

Figure 2.16 A simple feedback loop showing why ice melting and sea-level rise may accelerate

PPPPSS CONCEPTS

Think about the timescale over which future sea-level rise processes may take place. What are the main risks for people and places? Will individuals and societies in different places be able to adapt to these changes?

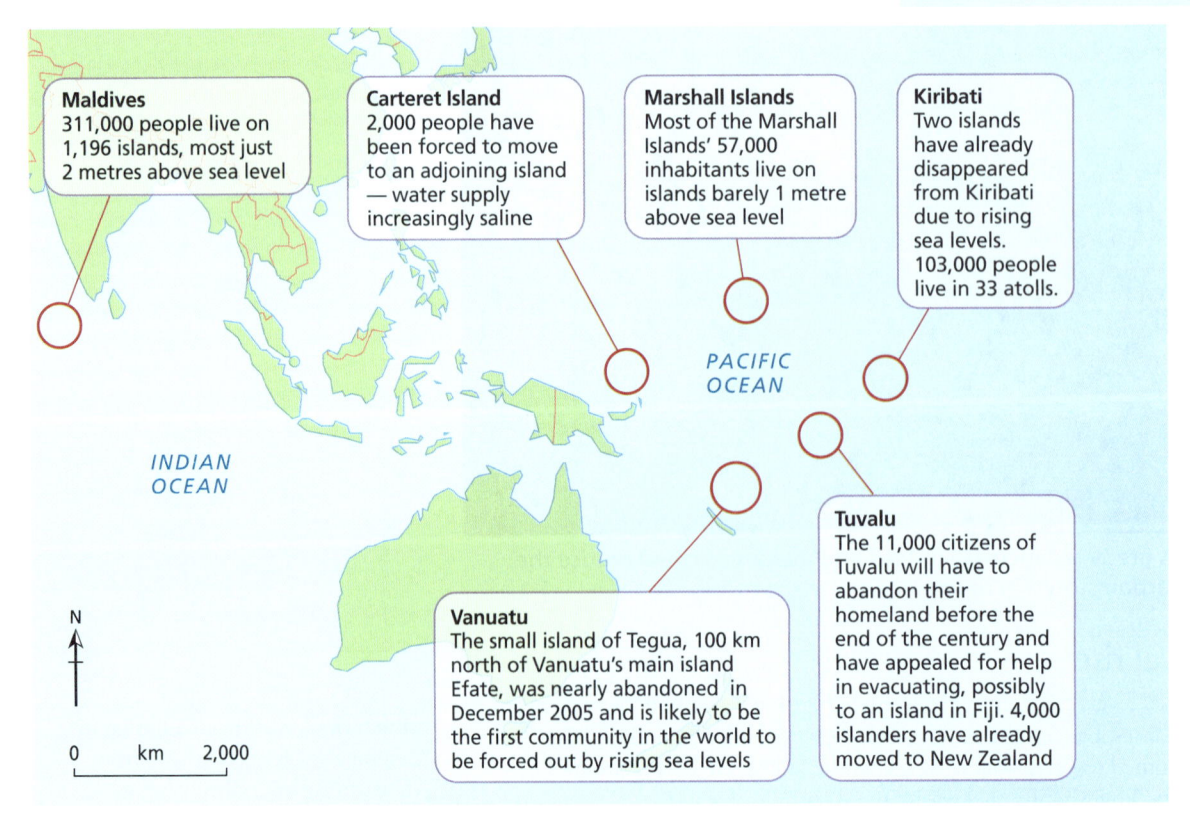

Maldives
311,000 people live on 1,196 islands, most just 2 metres above sea level

Carteret Island
2,000 people have been forced to move to an adjoining island — water supply increasingly saline

Marshall Islands
Most of the Marshall Islands' 57,000 inhabitants live on islands barely 1 metre above sea level

Kiribati
Two islands have already disappeared from Kiribati due to rising sea levels. 103,000 people live in 33 atolls.

Vanuatu
The small island of Tegua, 100 km north of Vanuatu's main island Efate, was nearly abandoned in December 2005 and is likely to be the first community in the world to be forced out by rising sea levels

Tuvalu
The 11,000 citizens of Tuvalu will have to abandon their homeland before the end of the century and have appealed for help in evacuating, possibly to an island in Fiji. 4,000 islanders have already moved to New Zealand

Source: Sue Warn

Figure 2.17 Islands threatened by rising sea levels

Extreme weather risks

An **extreme weather event** is defined as the occurrence of a value of a weather or climate variable (such as precipitation, wind strength or temperature) that is above or below a threshold value near to the previously observed maximum or minimum value. By definition, 'rare' high-intensity events *ought to* occur naturally from time to time. How, therefore, can we know if a given extreme weather event is (1) a rare occurrence that lies in the far tail of a statistical distribution but which was to be expected; or (2) an unexpected and unpredictable product of a changing climate?

Some leading scientists believe they can answer this question: they have concluded that global warming has 'significantly' increased the probability of some unusual recently occurring weather. The *State of the Climate* report published in 2012 by the Bulletin of the American Meteorological Society (AMS) showed a new emerging consensus among leading meteorologists. While they *cannot* say a particular event was or was not caused by climate change, the scientists argue that they *can* explain how the odds of such events have changed in response to global warming. Particular findings were that:

- the devastating Texan heatwave that occurred in 2011 is now *about 20 times more likely* during La Niña years (part of the ENSO cycle explained on page 46) than it was in the 1960s.
- the UK's exceptionally warm November 2011 temperatures are now *about 60 times more likely to occur* than they were in the 1960s.

Peter Stott of the UK's Met Office has argued that: 'It's like loaded dice. The chances of throwing a six have gone up a lot.'

Extreme weather events need looking at on a case-by-case basis, of course. The AMS scientists did not discover a clear human influence on any individual extreme event that they had studied. Natural variability will always bring periodic extreme floods, droughts and heatwaves to different places; it takes years of data to distinguish this from any underlying trend. For this reason, the precise link between various extreme weather events and climate change is still debated. However, as Table 2.5 shows, there are highly logical reasons why we should expect some hydro-meteorological hazards to become more commonplace on account of a rising GMST.

> **Keyword definition**
>
> **Extreme weather event** An occurrence such as drought or a storm which appears unusually severe or long lasting and whose magnitude lies at the extreme range of what has been recorded in the past.

> **PPPPSS CONCEPTS**
>
> Think about why there is uncertainty over whether particular extreme events can be attributed to climate change processes.

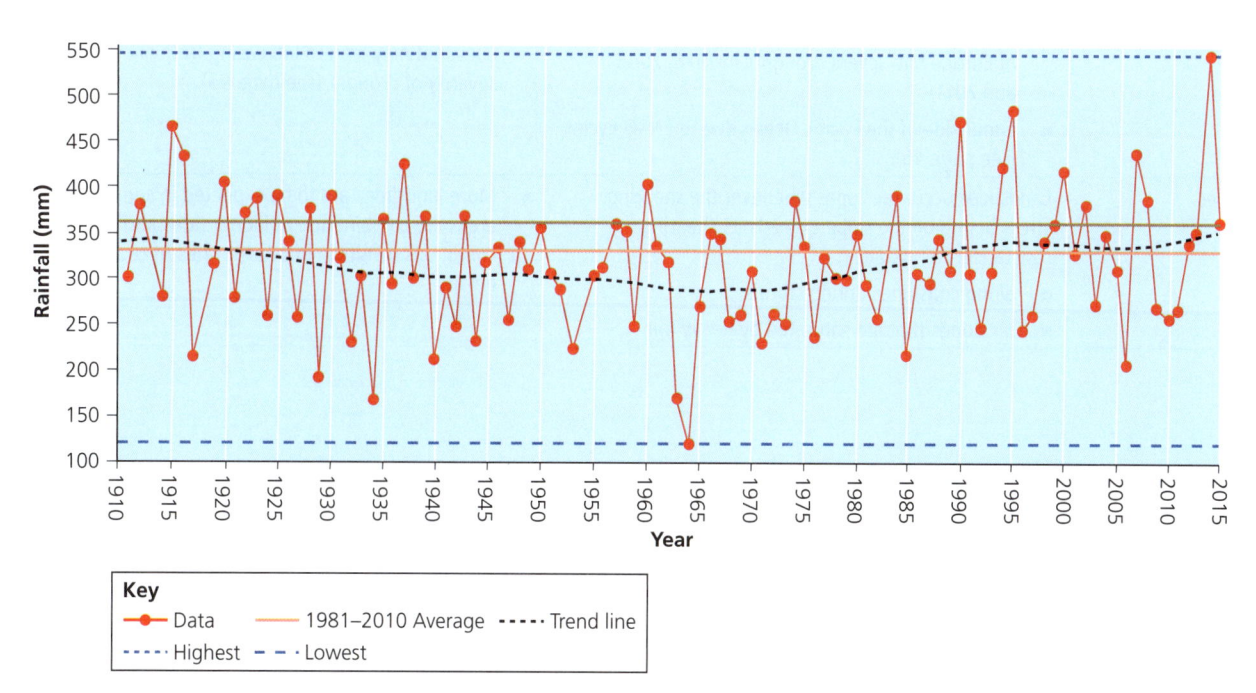

Figure 2.18 Winter rainfall in the UK, 1910–2015

Table 2.5 How climate change could increase the risk of hydro-meteorological hazards associated with extreme weather events

Type of hazard	Characteristics of this hydro-meteorological hazard	Why climate change might increase this hazard's strength or frequency
Floods	Floods occur when: ■ bankfull discharge is reached in a river ■ high tides drive seawater inland ■ water collects locally in depressions owing to high-intensity rainfall. The global pattern of flooding is complex but includes high occurrence: ■ along coastlines and on the floodplains of major continental rivers ■ in areas with a monsoon climate, especially deforested regions.	■ In a warmer world, more evaporation takes place over the oceans – and what goes up must ultimately come down. ■ Climate scientists believe rainfall patterns are changing in many parts of the world as the world's oceans warm. ■ Figure 2.18 shows UK winter rainfall between 1910 and 2015. One interpretation of these data is that there has been an increase in extreme winter rainfall since the 1980s.
Cyclones (hurricanes, typhoons and depressions)	Cyclones, or storms, are low-pressure weather systems that bring multiple risks of high wind speeds, heavy rainfall and flooding: ■ Most tropical cyclone (hurricane and typhoon) tracks lie 5–20° north or south of the equator (Figure 2.19). These tropical storms start life as a body of warm, moist air over a tropical ocean that has reached the critical temperature of 26°C. Driven by the rotating Earth's Coriolis force, a flow of air develops around a central eye. ■ Mid-latitude depressions form between 40°N and 60°N (where tropical and polar air masses converge at the polar front, under conditions of divergent flow in the jet stream). 'Bomb' depressions are exceptionally severe.	■ The devastation of New Orleans by Hurricane Katrina in 2005 ignited the important debate about whether climate change may be leading to the development of more high-intensity storms. ■ Cyclones draw strength from heat in ocean surface waters; basic physics and modelling studies suggest that tropical storms will become more intense, because warmer oceans provide more energy that can be converted into cyclone wind. ■ Some studies have shown more tropical storms occurring with a maximum wind speed exceeding 210 km per hour (category 4 and 5 storms on the Saffir–Simpson scale).
Droughts	A drought is an extended period of low rainfall relative to the expected average for a region. Drought periodically occurs: ■ in semi-arid or Mediterranean climates, when the rainy season is late ■ in the mid-latitudes, due to changes in the behaviour of air masses and the jet stream (parts of Europe had severe droughts in 1976, 2006 and 2011) ■ either side of the Pacific Ocean due to ENSO cycles (see page 46).	■ Europe's heatwave of 2003 led to 35,000 deaths, many of them elderly. This sparked debate on whether events that might previously have occurred once every 200 years might begin to recur every 20 or 30 years. ■ In IPCC climate change scenarios, many parts of the world are projected to experience temperature rises that may lead to increasing frequency and severity of drought (see page 95).
Landslides	Landslides occur on slopes whenever the shearing stress acting on the slope exceeds its shearing resistance. For instance: ■ along coastlines during storms ■ on slopes that are saturated by heavy rain.	■ More landslides are to be expected in some coastal areas if a combination of higher tides and stormier conditions results in increased rates of coastal erosion.

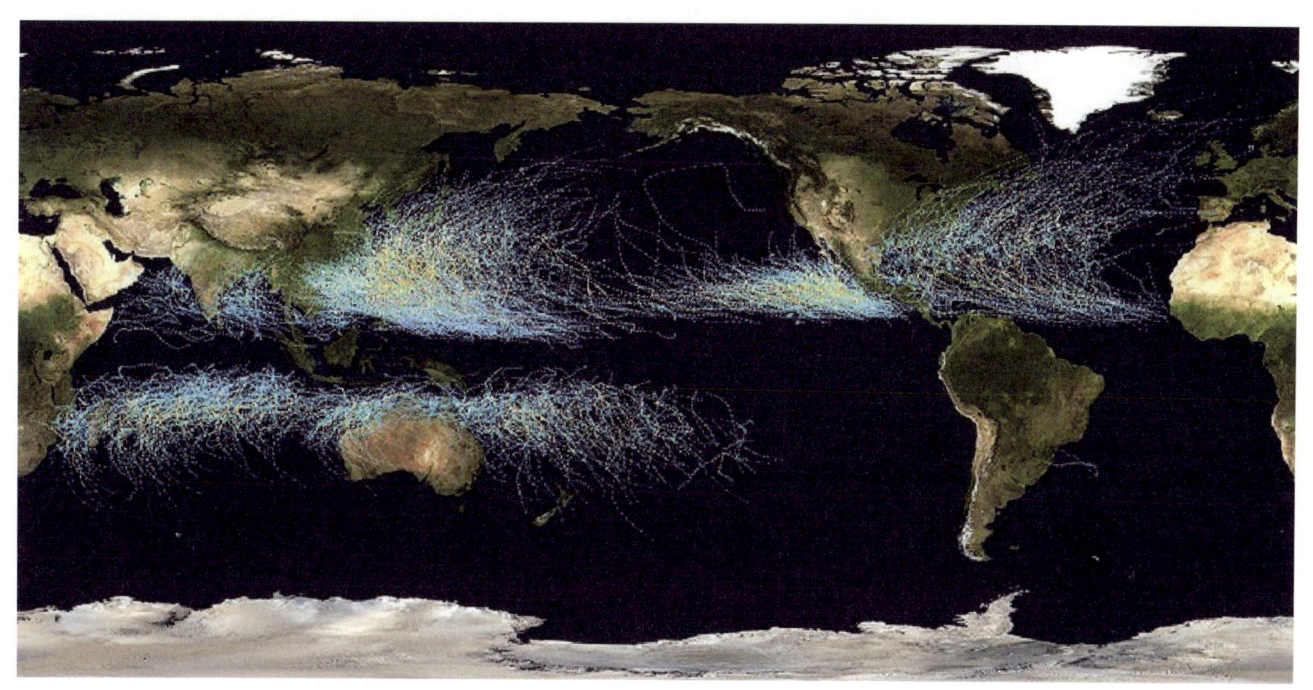

Figure 2.19 The tracks and intensity of tropical cyclones: how might more energy in the climate system impact on this pattern?

Spatial changes in biomes and ecosystems

Revised

Expected ecological changes linked with climate change include the longitudinal shift of vegetation **biomes**. Figure 2.20 shows the global biome map. By 2100, some of the boundaries may have shifted significantly. This is because, as the illustration suggests, the characteristics of the vegetation are determined by the climate currently found in different places.

- The boundary between coniferous forest and the polar **tundra** ecosystem in high latitudes of the northern hemisphere is called the **tree line**. Climate change research from the Scott Polar institute has reported that the tree line is definitely moving north on average. In places, the movement may equate to around 100 metres per year. The rate of movement is uneven, however, on account of local factors, including suitable soil and the absence or presence of animals that destroy saplings.
- The tundra biome is predicted to shrink in size by at least 20 per cent this century on account of the northward movement of trees (Figure 2.21) and the thawing of permafrost.

Keyword definitions

Biome Large planetary-scale plant and animal communities covering vast areas of the Earth's continents. For example, tropical rainforest, desert and grassland.

Tundra A 'cold desert' ecosystem composed of tough short grasses that survive in extremely cold, sometimes waterlogged, conditions at high latitudes where trees cannot grow. The tundra is underlain by permafrost.

Tree line The boundary between the coniferous forest and tundra biomes.

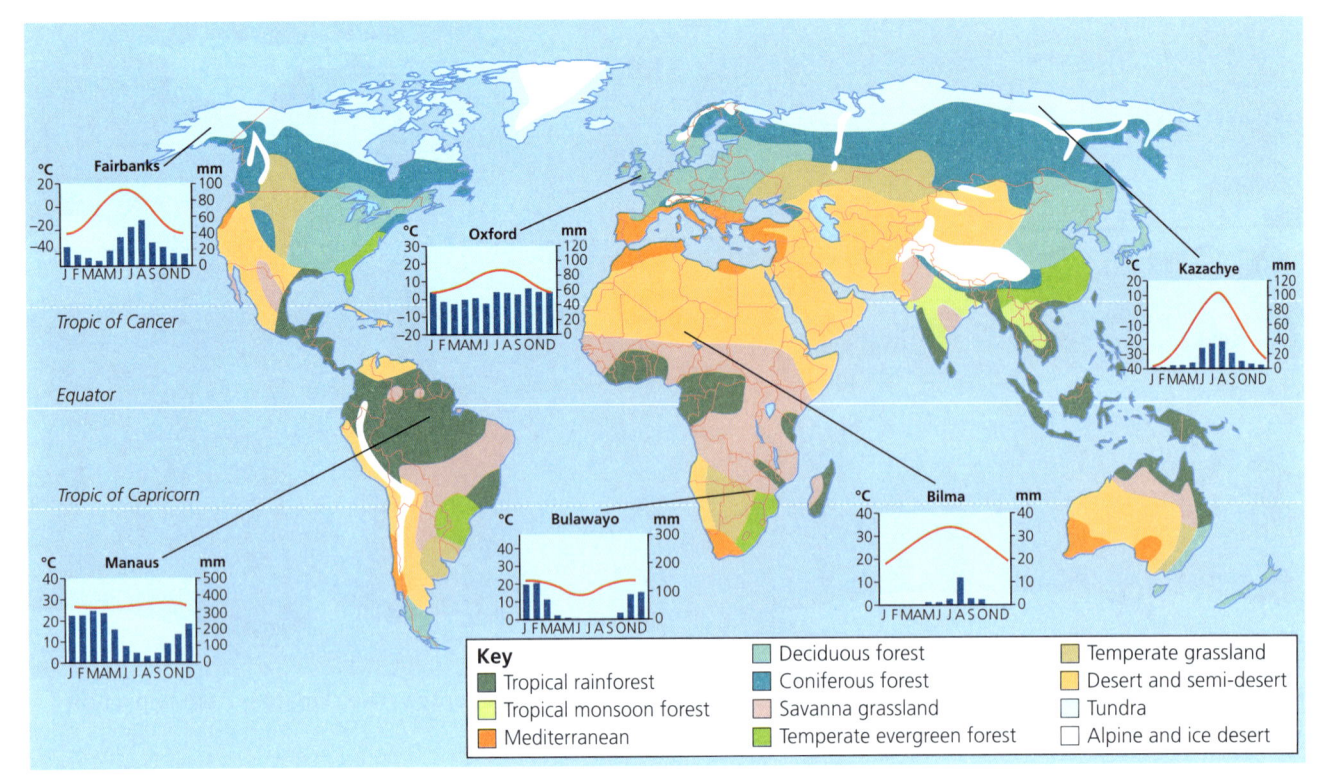

Figure 2.20 The global biome map

Figure 2.21 The current tree line in northern Russia; in the future, will it advance further north?

> **PPPPSS CONCEPTS**
>
> Think carefully about what the pattern of change is likely to be for Earth's biomes. Local factors in different places will influence what happens and not just climate change.

■ Habitat loss and animal migrations

Major changes to food webs are likely to occur in polar regions due to the larger temperature rises expected there.

- On land, many species will need to migrate northwards permanently, mirroring expected shifts in biome distribution patterns.
- Annual animal migration patterns will also be modified. Canada Geese will not need to fly as far south during winter; currently, around 90,000 of these birds spend the winter in UK estuaries and saltmarshes. In the future, they may not need to travel as far south from their Artic breeding grounds.
- Much of the polar bear's time is spent hunting for seals along the northern edges of the Arctic Ocean, in places where there is very little plant life. When seals are scarce, the polar bears roam inland into the tundra, in the very far north of Russia and Canada. In contrast, brown bears graze and hunt at the southern edges of the tundra, close to the forest boundary. To escape freezing conditions and food scarcity in winter, they hibernate, insulated by fat from the cold. In a rapidly warming scenario for the Arctic, many aspects of bear behaviour may change.
- Global warming will increase forest fires and insect-caused tree death in Arctic regions. The loss of old-growth forest represents a potential habitat loss for lichens, mosses, fungi and birds such as woodpeckers.

- Ocean life will be affected as Arctic ice thins and shrinks, and the water below warms. Summer ice has already decreased by about 10 per cent since 1979. There was a slight recovery in 2007–08 but the underlying trend is loss.

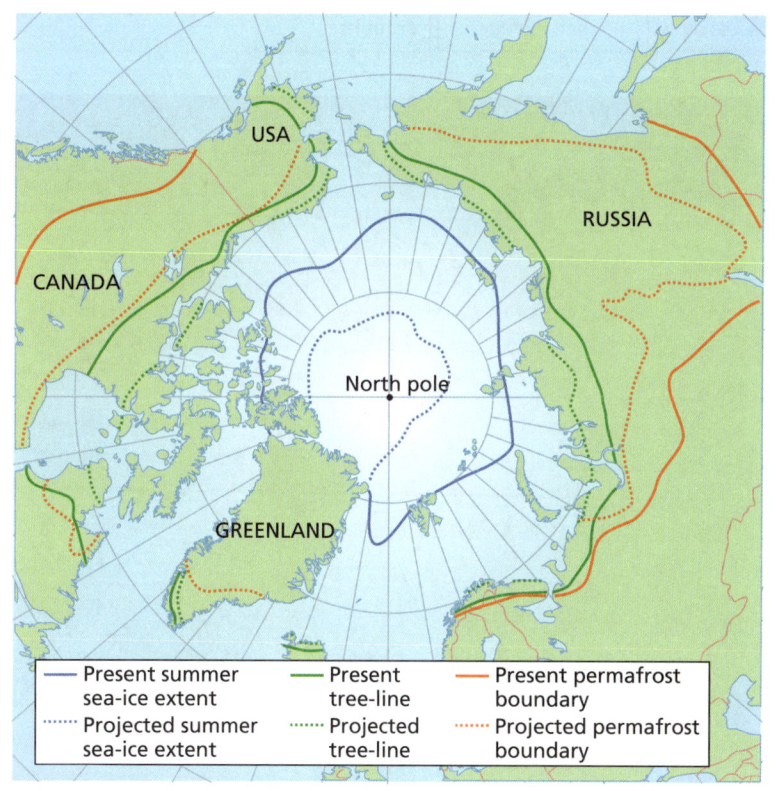

Figure 2.22 Arctic regional changes in sea ice cover, permafrost boundaries and the tree line (this assumes limited action to curb carbon emissions)

The human impacts and risks of global climate change

Revised

The future effects of climate change are by no means certain, as we have seen, but the potential risks are enormous. Risk is a key concept in the study of geography and can be defined as 'a real or perceived threat against any aspect of social or economic life, or the environment'. Figure 2.23 shows the risk equation: this is a way of exploring the concept of risk, which involves highlighting the role that a society's vulnerability and capacity to cope (or resilience) play in determining the size of a threat.

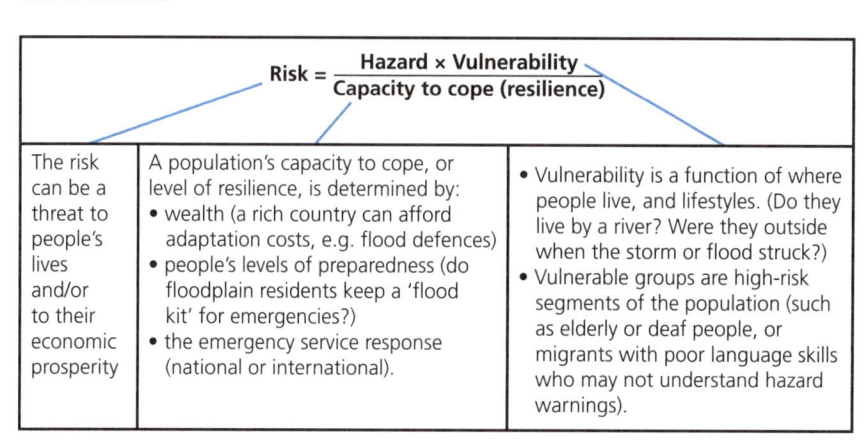

$$\text{Risk} = \frac{\text{Hazard} \times \text{Vulnerability}}{\text{Capacity to cope (resilience)}}$$

| The risk can be a threat to people's lives and/or to their economic prosperity | A population's capacity to cope, or level of resilience, is determined by:
• wealth (a rich country can afford adaptation costs, e.g. flood defences)
• people's levels of preparedness (do floodplain residents keep a 'flood kit' for emergencies?)
• the emergency service response (national or international). | • Vulnerability is a function of where people live, and lifestyles. (Do they live by a river? Were they outside when the storm or flood struck?)
• Vulnerable groups are high-risk segments of the population (such as elderly or deaf people, or migrants with poor language skills who may not understand hazard warnings). |

Figure 2.23 The risk equation

As we have seen, climate change is projected to increase the strength, frequency or unpredictability of extreme weather events. Figure 2.24 shows selected expected global effects of climate change and some of the associated risks for individuals and societies in different places.

Unit 2 Global climate – vulnerability and resilience

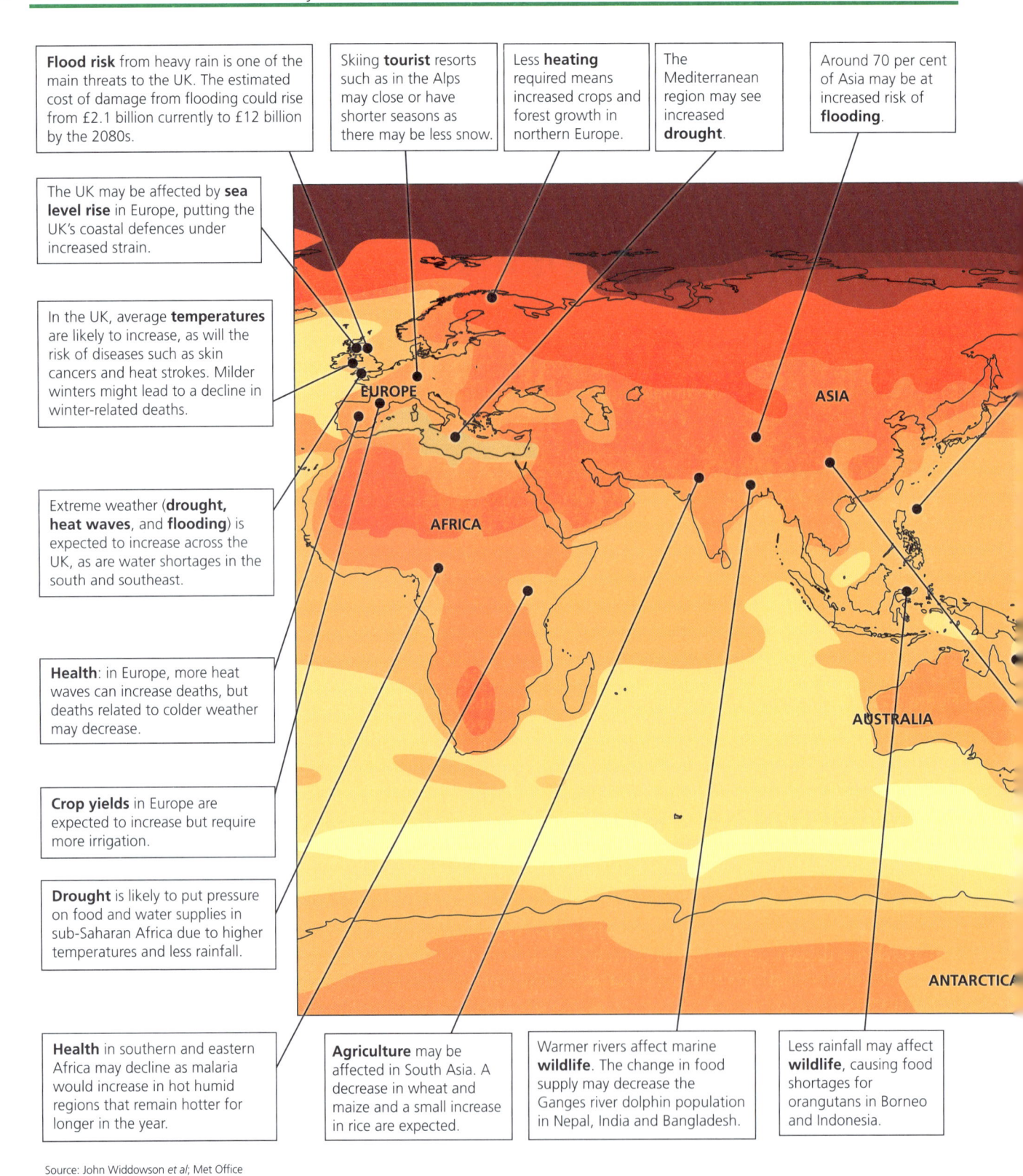

Flood risk from heavy rain is one of the main threats to the UK. The estimated cost of damage from flooding could rise from £2.1 billion currently to £12 billion by the 2080s.

Skiing **tourist** resorts such as in the Alps may close or have shorter seasons as there may be less snow.

Less **heating** required means increased crops and forest growth in northern Europe.

The Mediterranean region may see increased **drought**.

Around 70 per cent of Asia may be at increased risk of **flooding**.

The UK may be affected by **sea level rise** in Europe, putting the UK's coastal defences under increased strain.

In the UK, average **temperatures** are likely to increase, as will the risk of diseases such as skin cancers and heat strokes. Milder winters might lead to a decline in winter-related deaths.

Extreme weather (**drought**, **heat waves**, and **flooding**) is expected to increase across the UK, as are water shortages in the south and southeast.

Health: in Europe, more heat waves can increase deaths, but deaths related to colder weather may decrease.

Crop yields in Europe are expected to increase but require more irrigation.

Drought is likely to put pressure on food and water supplies in sub-Saharan Africa due to higher temperatures and less rainfall.

Health in southern and eastern Africa may decline as malaria would increase in hot humid regions that remain hotter for longer in the year.

Agriculture may be affected in South Asia. A decrease in wheat and maize and a small increase in rice are expected.

Warmer rivers affect marine **wildlife**. The change in food supply may decrease the Ganges river dolphin population in Nepal, India and Bangladesh.

Less rainfall may affect **wildlife**, causing food shortages for orangutans in Borneo and Indonesia.

Source: John Widdowson *et al*; Met Office

Figure 2.24 Projected global effects of climate change by 2100 (assuming 'business as usual' GHG emissions resulting in an average change of 3°C)

Impacts on agriculture, soils and crop yields

An important aspect of climate change to consider is the potential impact on agriculture. One third of the world's population earns a living farming – and the other two-thirds depend on the food these farmers produce.

Climate may change in ways that threaten both subsistence agriculture (the informal agricultural economy) and cash crops produced by the world's agribusinesses (large transnational farming and/or food production companies, such as Cargill or Del Monte).

2.2 Consequences of global climate change

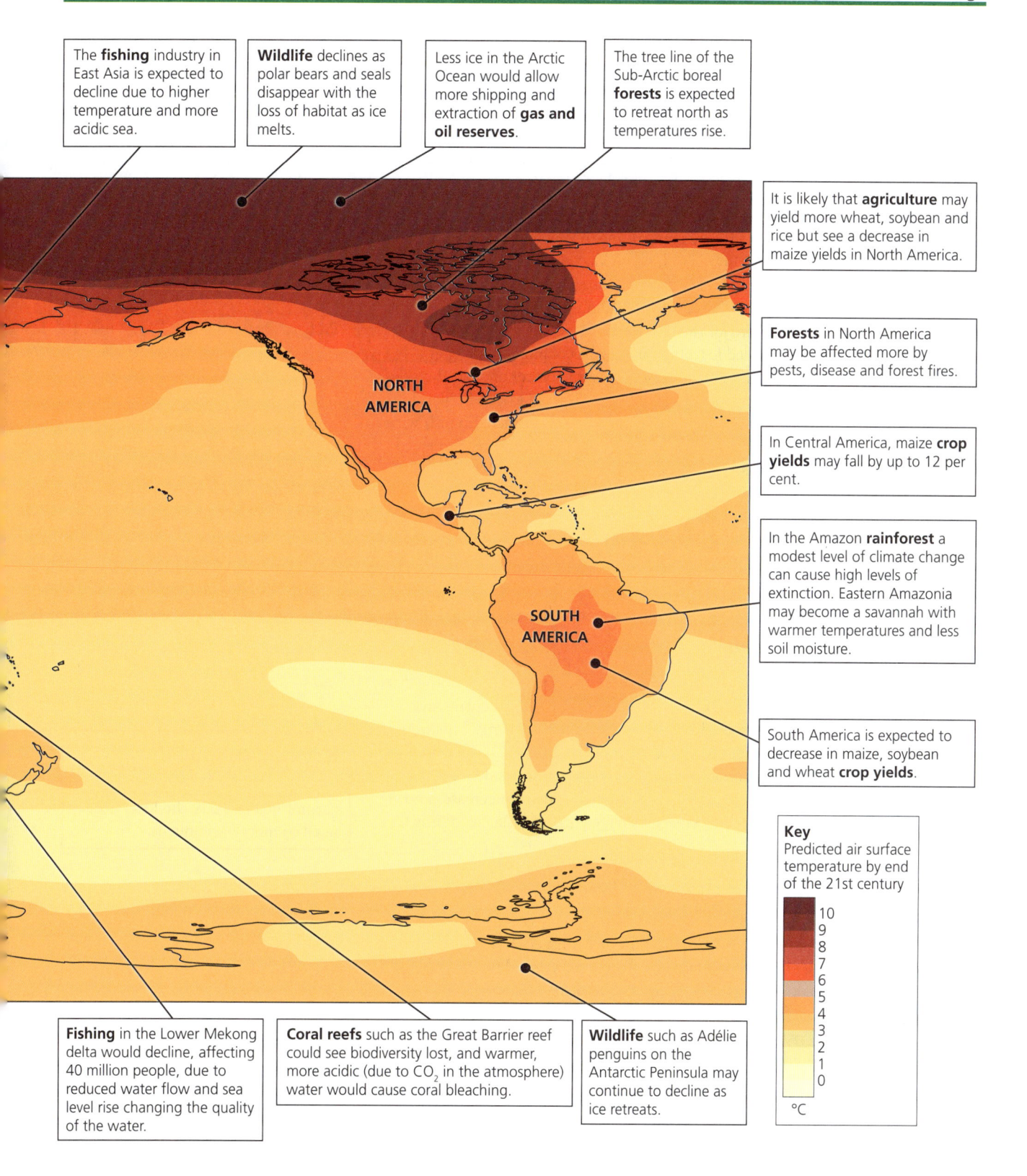

The **fishing** industry in East Asia is expected to decline due to higher temperature and more acidic sea.

Wildlife declines as polar bears and seals disappear with the loss of habitat as ice melts.

Less ice in the Arctic Ocean would allow more shipping and extraction of **gas and oil reserves**.

The tree line of the Sub-Arctic boreal **forests** is expected to retreat north as temperatures rise.

It is likely that **agriculture** may yield more wheat, soybean and rice but see a decrease in maize yields in North America.

Forests in North America may be affected more by pests, disease and forest fires.

In Central America, maize **crop yields** may fall by up to 12 per cent.

In the Amazon **rainforest** a modest level of climate change can cause high levels of extinction. Eastern Amazonia may become a savannah with warmer temperatures and less soil moisture.

South America is expected to decrease in maize, soybean and wheat **crop yields**.

Key
Predicted air surface temperature by end of the 21st century
10
9
8
7
6
5
4
3
2
1
0
°C

Fishing in the Lower Mekong delta would decline, affecting 40 million people, due to reduced water flow and sea level rise changing the quality of the water.

Coral reefs such as the Great Barrier reef could see biodiversity lost, and warmer, more acidic (due to CO_2 in the atmosphere) water would cause coral bleaching.

Wildlife such as Adélie penguins on the Antarctic Peninsula may continue to decline as ice retreats.

- Desertification and the extension of arid conditions in sub-Saharan Africa could adversely affect agriculture in regions like Same District in Tanzania and Kitui District in Kenya.
- Soil erosion by water is a major threat to the sustainability of agriculture in many tropical and subtropical regions in both low-income and high-income countries. Global warming is expected to lead to a more vigorous hydrological cycle, including more total rainfall and more frequent high-intensity rainfall events. These rainfall changes could have significant impacts on soil erosion rates. Some research suggests rainfall erosivity levels may be on the rise already

across parts of the USA. According to soil scientists, when rainfall amounts increase, erosion and runoff will increase at an even greater rate: the ratio of annual rainfall increase to erosion increase is in the order of 1 : 7.
- Productive land could be lost to sea-level rise, especially valuable fertile alluvial soils on river floodplains and deltas.
- Increased hurricane frequency could devastate agriculture (Hurricane Katrina cost the USA US$1 billion in crop and poultry losses).

Not all climate change impacts on soil and agriculture are negative, of course. Shifting rainfall, temperature and biome patterns mean that certain areas will become more productive or better suited to higher-value crops or land uses.
- Positive impacts on the limits of cultivation can already be seen in southern Greenland where the growing season is a month longer than it used to be. In 2007, broccoli was grown there for the first time.
- The climatic conditions needed for different varieties of grape to prosper are highly specific. As a result of climate change, some places are gaining newfound success as wine-growing regions. Vineyards in southern England have gone from strength to strength in recent years, for instance. English wine producers attribute their newfound success to warmer temperatures produced by climate change (helped also by geological similarities between southern England and France's Champagne region).

Health hazards

Another aspect of climate change to consider is the health hazard posed by disease and food insecurity:
- Vector-borne diseases such as malaria and water-borne diseases (diarrhoea) could increase with climate change.
- Some scientists think that climate change may have played a role in ebola outbreaks by changing the behaviour of bats (the suspected carriers of the 2014 ebola outbreak). Computer modelling suggests that, in parts of central and western Africa, the range of some bat species could be expanding. This would mean increased contact between bats and humans.
- Some places may experience increasing **food insecurity** alongside water shortages: less food will be produced in regions where rainfall decreases and desertification occurs (such as the Sahel and large parts of Asia, according to IPCC projections).

> **Keyword definition**
> **Food insecurity** When people cannot grow or buy the food they need to meet basic needs.

Migration and transport movements

United Nations agencies estimate nearly 10 million people from Africa, south Asia and elsewhere have migrated or been displaced by environmental degradation, weather-related disasters and desertification in the last 20 years. The UN predicts a further 150 million people may have to move in the next 50 years and has identified 28 countries now at extreme risk from climate change. Of these, 22 are in Africa. The majority of people displaced by more severe climates will be the very poor. Unable to afford to travel long distances, they are likely to become IDPs or refugees in neighbouring states (see page 20).
- The International Red Cross estimates that each year 400 weather-related disasters result in forced migration. As we have learned, while climate change cannot be linked conclusively to the occurrence of any one particular event, there is an anticipated correlation between a higher GMST and an increased frequency of extreme weather.
- More people may become climate change refugees on account of rising sea levels. Dozens of islands in the Indian Sunderban region are being regularly flooded, threatening thousands. The Maldives, located southwest of India and Sri Lanka, is gravely at risk from rising sea levels. It consists of a chain of 1,190 low-lying islands that are surrounded by the waters of Indian Ocean. Bangladesh is the most vulnerable large country, with 60 per cent of its land less than 5 meters above sea level. Page 20 explored how coastal regions of Bangladesh are vulnerable to rising sea levels. Bhola villagers will have no option but to migrate; many will go to the slums of Dhaka.

Shipping movements in an ice-free Arctic

With summer sea ice currently shrinking by 12–13 per cent a decade, the Arctic Ocean could be ice free by 2050. Already, it is becoming more accessible for shipping:

- In 2017, new icebreaking liquid natural gas (LNG) carriers began to travel from Siberia to the Pacific. Russia is no longer wholly dependent on selling gas through pipelines to Europe.
- The new shipping vessels used to carry the gas have been engineered to withstand temperatures as low as minus –50°C.
- The route from Yamal, Siberia to Japan takes just 14 days during summer months via the new northern sea route along the Arctic coast of Russia (Figure 2.25).
- In winter, when sea ice reforms, the route closes. In the future, however, it may be possible for vessels to make the same journey all year round.

Figure 2.25 The new summer route taken by Russian shipping

Risk amplification

Climate change risks are heightened further by a range of human processes:

- Population growth can lead to increased numbers of people living in a vulnerable state of poverty. For instance, numbers living in sub-Saharan Africa have doubled since 1980, meaning that twice as many people are now at risk of famine when rains fail.
- Many large cities and megacities are places where predicted climate change risks are high. In particular, many more people will live in increasingly high-risk urban zones situated along floodplains and coastlines. Between 2010 and 2020, Lagos and Kinshasa (Democratic Republic of Congo), will add 3.5 million and 4 million people to their respective populations. Both have a known flood risk and much of the new housing growth will be precisely in areas where the risks are greatest (for instance, in areas of informal slum housing built around the lagoons or on coastal sandbars in Lagos).
- Deforestation throughout Southeast Asia (Bangladesh, Malaysia, India and Indonesia) has modified hydrological systems, making overland flow far more

likely. The biomass interception store has been removed (which naturally functions like a valve, slowing down the passage of rainwater onto the land). Inevitably, flooding increases in deforested areas like the Ganges basin.
- Deep-rooted conflicts, such as India and Pakistan's dispute over the Kashmir region, could be exacerbated by food and water shortages in the future, potentially jeopardizing international cooperation aimed at tackling climate change.

> **PPPPSS CONCEPTS**
> Think about the way human and physical processes combine to create the possibility of a riskier world, where more people may be forced to migrate for safety.

■ KNOWLEDGE CHECKLIST:
- The hydrosphere, atmosphere and biosphere, and their linkages
- Projected effects of climate change on the cryosphere (water stored as ice)
- Projected effects of climate change on global carbon stores (water, ice and the biosphere)
- The causes and possible consequences of global sea-level rise
- Evidence for increasing risks (frequency and severity) of different types of extreme weather
- Projected effects of climate change on biomes, the tree line, habitats and animal migration
- How climate change could affect agriculture and crop yields
- Climate change risks for human health (food availability and the incidence of disease)
- The implications of climate change for patterns of migration and ocean transport

EVALUATION, SYNTHESIS AND SKILLS (ESK) SUMMARY:
- How the evidence used to predict the timing and scale of possible climate change is complex, difficult to collect and sometimes contradictory
- How the impacts of climate change are likely to be uneven and to affect some places and people more adversely than others (some may even benefit)

EXAM FOCUS

COMPLETE THE ANSWER

Below is a sample AO2 short-answer question (based on Section A of the examination), and an answer to this question. The answer contains a basic point and an explanatory link back to the question, but lacks any development or exemplification. Without this, full marks are unlikely to be awarded. Write out both parts of the answer, then improve it by adding your own development or exemplification to the two basic reasons shown below.

Explain two possible effects of global climate change on the amount of the world's water which is stored in ice. [2 + 2 marks]

1. Warmer temperatures would lead to widespread melting of ice. For instance, large amounts of water are stored in glaciers around the world. Ice in these places will begin to melt.
2. The largest amounts of ice are stored in Antarctica and Greenland. Climate scientists worry that a high rise in global temperature may lead to this ice melting, which would reduce the amount of water stored in ice.

ELIMINATE IRRELEVANCE

Below is a sample AO2 short-answer question and an answer to this question. Read the student answer carefully and identify parts that are not directly relevant to the question asked. Note down the information that is irrelevant and justify your deletions.

Suggest two possible positive impacts of climate change for some people. [2 + 2 marks]

1. Rising global temperature may lead to a shift in global biomes, such as the tropical rainforest or Arctic tundra. The tree line is predicted to move northwards, meaning that more land in Russia and Northern Europe, along with Canada, will become forest. As a result, people will have to adapt to these changes.
2. Arctic sea ice cover has already begun to shrink at an alarming rate. At the end of 2016, air temperature in the Arctic was 20°C higher than is normal for the time of year. Once the ice begins to melt, changes in albedo mean that the rate of melting increases even more rapidly because of positive feedback. Polar bears lose valuable hunting habitats. However, one benefit may be that it is easier for shipping to make use of the Arctic region.

Avoid repetition

In each part of the first answer, only a very simple idea has been set out. But in both cases, there is repetition of this idea without adding any valuable extra explanation, development or exemplification. Try to avoid doing this. Instead, add some real detail of a place or environment, such as the name of a glacier or a fact about the volume of water stored in Antarctica.

2.3 Responding to climate change and building resilience

Any future temperature rise must be limited to less than 2°C above pre-industrial levels if irreversible, unmanageable and highly damaging impacts are to be avoided for the environment and societies. This equates to around 550 ppm of carbon dioxide in the atmosphere. Why, therefore, are the scale and pace of action to tackle climate change not greater? Why is more not being done to reduce carbon emission and to build **resilience** against environmental changes, some of which – according to the IPCC – are now unpreventable? The answer to these questions may lie in Table 2.6. It analyses disparities in exposure to climate risk and vulnerability among different groups of people.

- While some countries and groups of people have high risk exposure, others do not (at least in the short term).
- The power to act is also spread unevenly among stakeholder groups. Many of the people and countries who have most to lose from climate change also have least power to bring about the real political changes that are needed at both national and global levels (such as binding commitments to reduce carbon emissions and fossil fuel use, while also investing heavily in renewable energy).

> **Keyword definition**
>
> **Resilience** The capacity of individuals, societies, organizations or environments to recover and resume 'business as usual' functions and operations following a hazard event or other system shock.

Table 2.6 Disparities in the vulnerability of different people and places to climate change might affect their perception of risk

Characteristics of people and places that may affect vulnerability		How personal and societal perspectives might vary on the need to respond to climate change
Location and vulnerability	• Temperate or semi-arid region? • Coastline or continental interior? • High or low latitude?	People living where risks are greatest – including islands and coastal regions, areas at risk of increased drought and high latitudes – may want urgent action.
Wealth, education and vulnerability	• Low- or high-income countries? • Are citizens well-educated about climate change risk and resilience?	Well-educated citizens in high-income countries may know more about the issues; but they may also be more confident that they can adapt to climate change.
Age, gender and vulnerability	• Older people or younger people? • Men or women?	Younger people who will be alive in the 2070s and 2080s may want to take action to safeguard their own future. Men's and women's views tend to be broadly similar.
Risk perceptions and vulnerability	• What does government and the media say about climate change?	People may be influenced by politicians and news channels that are skeptical about climate change. US President Donald Trump is a skeptic.

> **PPPPSS CONCEPTS**
>
> When you have finished reading this chapter, review Table 2.6 and think about the extent to which you agree or disagree with its propositions.

CASE STUDY

CLIMATE CHANGE VULNERABILITY IN THE PHILIPPINES

The Philippines is a middle-income nation of 101 million people with a GDP per capita (PPP) of US$8,200 in 2017. In the last 20 years, around 60,000 people have lost their lives in this country to the combined effects of different natural disasters.

The 7,000 islands are located at latitudes 5–20°N of the equator – within the tropical cyclone (typhoon) belt (Figure 2.26). They are hit by around 20 major storms each year. Some of the islands are isolated, making it harder to warn people living there of approaching storms. In 2011, a combination of coastal and river flooding, driven by Typhoon Washi, washed away slum housing on the banks and sandbars of the Cagayan river, killing 1,250 people. On the island of Mindanao, half a million people lost their homes. In 2013, Typhoon Haiyan – the strongest storm ever recorded making landfall in the Philippines – killed 6,300 people (Figure 2.27). Some scientists think there is a correlation between the increasing intensity of storms striking the Philippines and the progression of climate change.

The human and economic impacts of storms can be severe and are rising over time because of:

- the vulnerability of poorer segments of the country's increasing population (many people live at sea level in poorly constructed homes; total population grew by 45 per cent between 1990 and 2008)
- failure of the authorities to have better management policies, such as land-use zoning, despite the known risks
- the growing value of property and business assets (Manila is a megacity of more than 20 million people where economic risks are very high: it is now a major call centre hub, for instance).

Figure 2.26 A major tropical storm (typhoon) passing over the Philippines

Figure 2.27 The impact of Typhoon Haiyan in the Philippines

CASE STUDY

CLIMATE CHANGE VULNERABILITY IN LONDON, UK

London faces a threat of flooding caused by storm surges driven by depressions into the Thames Estuary from the North Sea. The last serious event was in 1953 when 300 people died. The UK's Environment Agency believes there remains a one-in-a-thousand chance of London being flooded in any given year, owing to the limits of the protection offered since 1984 by the Thames Flood Barrier (Figure 2.28). Closing the barrier seals off part of the upper Thames from the sea and unusually high tides that might push seawater into central London. When not in use, the six rising gates rest out of sight on the riverbed, allowing free passage of river traffic though the openings between the piers.

However, climate change may mean that the risk of the barrier failing is growing because of (1) global eustatic sea-level rise (mainly because of thermal expansion) and (2) more powerful storms (see page 54) that generate higher tides (also known as storm surges). By 2030, either a new US$6 billion flood barrier or a US$30 billion tidal barrage may need to be built. The high cost of defending this city appears to be justifiable when risk and vulnerability are examined in detail:

- Eight million people live in London, around one million of whom (in half a million homes) are at direct risk of flooding. Elderly or disabled floodplain residents are especially vulnerable to a sudden-onset flood event.

- If the Thames did burst its banks in central London due to a tidal surge, Westminster (where the UK Government is based) would be under 2 metres of water; 16 hospitals and 400 schools would be flooded; 68 underground railway stations would be drowned utterly.

- London's total level of risk is growing all the time, as more people migrate there and new housing developments increase the total value of vulnerable property. An estimated US$200 billion of property is now at risk.

Such disastrous events are most likely preventable, however. London is protected already to a very high standard and the UK government can be expected to pay for whatever new adaptation measures are required as and when climate risks worsen.

Figure 2.28 The Thames Flood Barrier

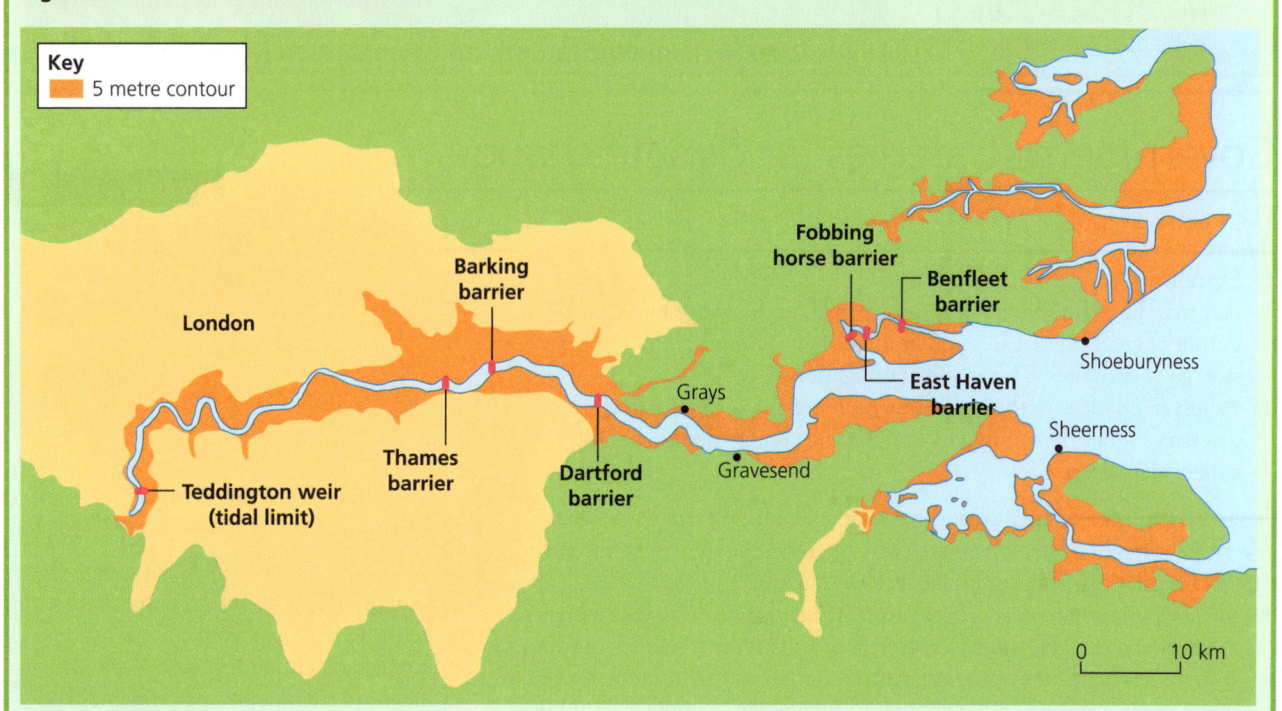

Figure 2.29 Parts of London at risk from a 5-metre storm surge (the Thames Barrier currently protects areas of central London)

CASE STUDY

CLIMATE CHANGE VULNERABILITY AMONG INDIGENOUS ARCTIC COMMUNITIES

Local indigenous populations in the Arctic region face an entirely unsustainable future on account of climate change, according to scientists who are working in the region. Life is changing irreversibly for people who live in the Arctic and depend on the presence of sea ice both economically and culturally.

- In northwest Greenland, many of the region's native Inuit maintain a strong cultural connection to the Arctic landscape through their traditional livelihoods of hunting and fishing. An important part of the traditional way of life is the seasonal hunt for different animal species, including seals and narwhals.

- In places where the sea ice is breaking up earlier and forming later, however, they are losing safe access to traditional hunting grounds for large parts of the year. In the future, there will be no access at all (Figure 2.30). Observations show that the Arctic is warming twice as fast as the rest of world because of its high latitude. Summer sea ice cover has decreased by about 50 per cent since the late 1970s (see page 57).

- On Greenland's Disko Island, the main town of Qeqertarsuaq is home to around 900 inhabitants. Thick winter sea ice used to provide dog-sledding routes that connected Disko other places. The recent reduction in sea ice means that Disko residents are now becoming isolated, in addition to losing their ice hunting grounds.

This is an extreme form of vulnerability: Inuit culture and the physical environment are so highly interconnected that it is hard to imagine how communities will survive in any recognizable way if the ice does not.

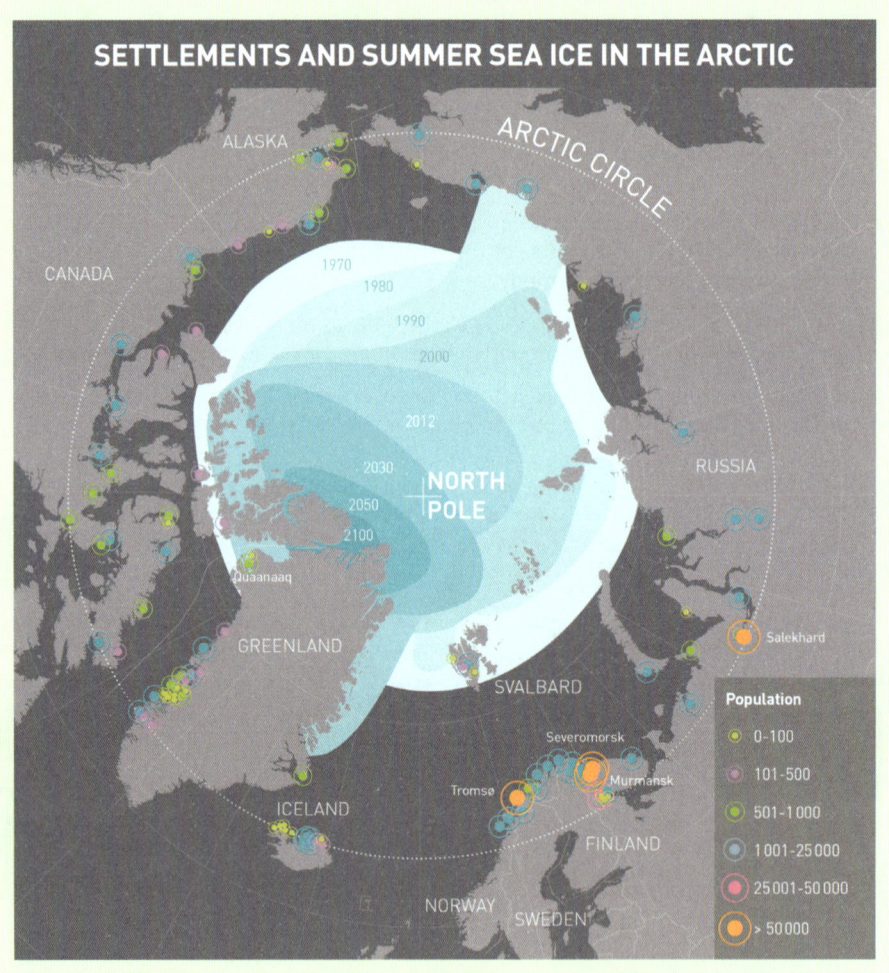

Figure 2.30 Settlement positions and summer sea ice in the Arctic

Government strategies for climate change

Revised

There are three broad ways in which governments respond to the threat of climate change:
- Do nothing.
- Act to reduce the severity or intensity of climate change by reducing the output of greenhouse gases and/or increasing the size and amount of GHG storage or sink sites. This is called **mitigation**.
- Adjust to changes in the environment, for example by building coastal defences, rather than trying to stop climate change from happening. This is called **adaptation**.

Figure 2.31 shows examples of mitigation and adaptation strategies which a government might adopt. There are two important differences between the two approaches that can affect political decision making:
- Mitigation has *upfront* economic costs. Humans need to spend money now to develop renewable sources, recycle more and reduce energy consumption.
- Adaptation has *future* economic costs. Humans in the future will need to spend money to cope with a changing climate, for example by building higher flood defences or growing drought-resistant crops.

Keyword definitions

Mitigation Any action intended to reduce GHG emissions, such as using less fossil fuel-derived energy, thereby helping to slow down and ultimately stop climate change. Mitigation can be practised by stakeholders at different scales, from a citizen switching off a light, to a government setting strict national targets for reduced carbon emissions.

Adaptation Any action designed to protect people from the harmful impacts of climate change but without tackling the underlying problem of rising GHG emissions.

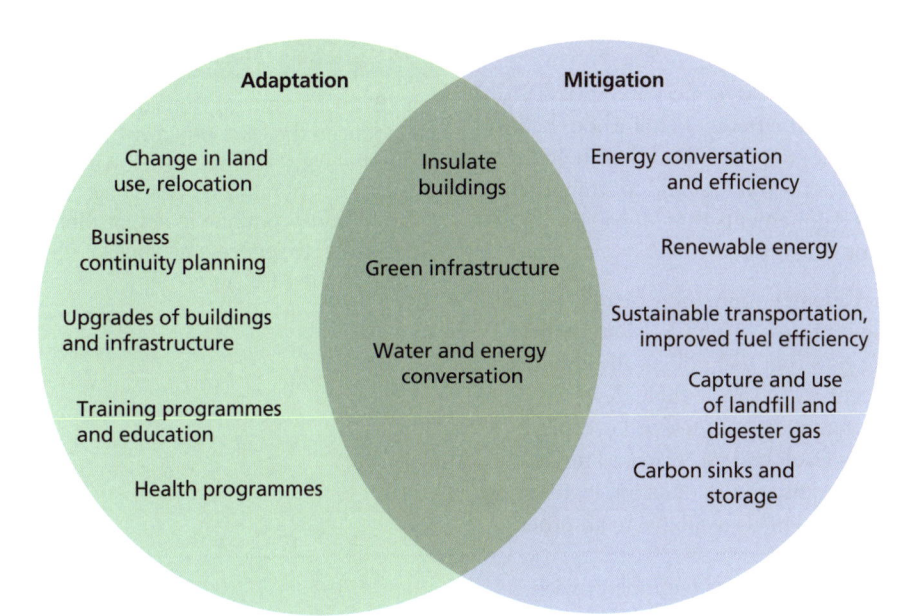

2.3 Responding to climate change and building resilience

Adaptation
- Change in land use, relocation
- Business continuity planning
- Upgrades of buildings and infrastructure
- Training programmes and education
- Health programmes

Overlap (Adaptation & Mitigation)
- Insulate buildings
- Green infrastructure
- Water and energy conversation

Mitigation
- Energy conversation and efficiency
- Renewable energy
- Sustainable transportation, improved fuel efficiency
- Capture and use of landfill and digester gas
- Carbon sinks and storage

Figure 2.31 Adaptation and mitigation strategies a government might adopt
Source: David Redfern

PPPPSS CONCEPTS
Think about the extent to which the local and national governments in the place where you live have adopted either mitigation or adaptation strategies for climate change.

■ Global geopolitical efforts

Global concern about climate change has been mounting since the late 1980s. Figure 2.32 summarizes how the global community has moved forwards slowly over a 30-year period towards a shared agreement on how to tackle the issues. Events can be studied in greater detail online using the UN's interactive climate change action timeline: **http://unfccc.int/timeline/**

1988
- In 1988 the UN Environmental Programme and the World Meteorological Organization set up the Intergovernmental Panel on Climate Change (IPCC).
- The IPCC began detailed research into the enhanced greenhouse effect and the role played by human activities in driving up atmospheric levels of carbon dioxide, methane and nitrous oxide.

1992
- Key political players first began to discuss their official response to IPCC findings at the 1992 Earth Summit in Rio.
- 190 countries signed a treaty agreeing that the world community should 'achieve stabilization of greenhouse gas concentrations in the atmosphere at a low enough level to prevent dangerous anthropogenic interference with the climate system'.

1997
- World leaders met in Kyoto in order to develop the treaty further into a more detailed binding agreement known as a protocol.
- The resulting Kyoto Protocol required all signatories to agree to a legally binding GHG emissions reduction target. However, the effectiveness of the Kyoto Protocol was weakened when some countries, notably including the USA, chose not to sign it, fearing the cost of emission reductions (either through paying for new green technology or reducing industrial output).

2009
- Successive UN meetings in Copenhagen (2009), Cancún (2010) and Durban (2011) focused on the need for all nations to make a legally binding pledge. However, these conferences failed to arrive at a legally binding agreement.
- Rich countries agreed in principle at Copenhagen to set up a US$100bn 'green fund' for poorer countries by 2020 to assist with adaptation and mitigation efforts.

2015
- The UN negotiated a new international climate change agreement for all countries at the 2015 Paris Climate Conference, also known as COP21.
- The Paris Agreement commits 195 countries to reducing their emissions of greenhouse gases so that the future average atmospheric temperature will not exceed 2°C above pre-industrial levels.

Figure 2.32 A timeline for global geopolitical efforts to tackle the challenge of climate change

Strengths and weaknesses of COP21

World political leaders officially signed up to the Paris Agreement in April 2016. As of February 2017, 195 UN members had signed the treaty, 132 of whom had ratified it. The Paris Agreement requires all parties to put forward their best efforts through 'nationally determined contributions' (NDCs) and to strengthen these efforts in the years ahead. This includes requirements that all parties report regularly on their emissions and on implementation efforts.

The Paris Agreement's central aim is to avoid a temperature increase of 2°C, and make every effort to limit the increase to a lower target of 1.5°C. Additionally:
- GHG emissions will be allowed to rise for now (but technologically enhanced carbon capture will be needed later this century to reduce GHG levels).
- Emissions targets will be set by countries separately, but reviewed every five years. After each five-year review, emissions levels will be decreased further.
- The Agreement also provides for enhanced transparency of action by insisting that accurate emissions records will be kept and made available to all other countries.
- Wealthy countries will share their low-GHG science and technology with developing countries and the most vulnerable countries.
- Wealthy countries will provide finance for low-income nations most affected by anthropogenic climate change.
- Developed countries that have historically contributed a large proportion of the anthropogenic GHG stock in the atmosphere will recognize the 'loss and damage' inflicted on poor countries because of climate change. In doing so, they acknowledge that the sources of GHG emissions may be spatially distant from the countries most impacted by them.

Some strengths and weaknesses of the outcomes from COP21 Paris Agreement are shown in Table 2.7.

> **PPPPSS CONCEPTS**
>
> In 2017, US President Donald Trump announced the USA would withdraw from the Paris Agreement. However, the Mayors of many US cities have said they will continue to act, at the local scale, to reduce emissions in line with the Paris Agreement.

Table 2.7 Strengths and weaknesses of the COP21 Paris Agreement

Strengths	Weaknesses
- Getting 195 countries to agree on anything is a major achievement! - From a scientific perspective, COP21 gives us hope that so-called 'dangerous climate change' across all continents can be avoided. - From an economic perspective, the agreement gives countries time to decarbonize their economies 'without sacrificing economic prosperity on the altar of environmental wellbeing'. - From a political perspective, it allows all governments to hold each other to frequent account regarding emissions levels and targets. Ten-year reviews, for instance, would be too infrequent. - Finally, from the perspective of poorer, low-lying countries like Bangladesh, elements of COP21 'promise a degree of justice for those adversely affected by wealthier countries' previous GHG emissions'.	- Massive GHG reductions will be required by 2050 to keep the temperature increase below 2°C; some countries may find it too expensive to phase out things like coal-fired power stations over the required timescale. - If a country is hit by an economic recession or experiences political change, its priorities may alter and climate could slip down the domestic agenda. (In the USA, President Trump has reversed decisions made by President Obama was, for instance.) - Replacing fossil-fuel economics with renewable-energy ones sufficient to maintain decent lifestyles requires technologies that have yet to be invented. - The Paris Agreement is 'merely a statement of intent, albeit an important one. The key to its success or failure lies in…the fine details of how countries respond when one or more fail to honour their commitments'.

Source: Noel Castree, *Geography Review* **30** (1)

■ European agreements

In addition to UN action on climate change, European Union (EU) member states have agreed among themselves to curb carbon emissions.
- The European Union is the world's largest political grouping of high-income nations.
- EU countries acknowledge that they are responsible for much of the anthropogenic carbon stock already in Earth's atmosphere (as opposed to the current carbon flow that also comes from a large number of developing countries and emerging economies).
- Accordingly, the EU has pledged to cut emissions by 20 per cent by 2020 (compared with 1990 levels). By 2050, EU states have agreed to cut their emissions by 80 per cent (by making continued progress towards a 'low-carbon society').

> **PPPPSS CONCEPTS**
>
> Think about why a binding global agreement on climate change is thought to be essential if limiting global temperature increase to 2°C is to remain a possibility.

■ Carbon emissions trading and offsetting

Carbon emissions trading (CET) schemes are an example of climate change mitigation. The aim is to achieve a progressive reduction in the total of GHG emissions produced by a national or regional economy.
- CET schemes are a form of environmental economics: this is viewed by some economists as the best way of protecting the world from environmental harm while avoiding excessive amounts of direct government regulation. Instead, market forces are relied on to bring about the changes that are needed. CET schemes use a market solution to tackle an environmental problem.
- The idea is to harness the market power of businesses and consumers. Both groups can be expected to choose low-carbon products and services if they are priced more expensively than high-carbon goods and services.
- CET schemes operate using a **cap and trade** system.

> **Keyword definition**
>
> **Cap and trade** An environmental policy that places a limit on the amount a natural resource can be used, identifies the resource users, divides this amount up into shares per user, and allows users to sell their shares if they do not wish to use them directly.

■ The EU emissions trading system (EUETS)

EUETS is the world's largest CET scheme. Introduced by their European Union in 2003, it encompasses thousands of power stations, factories, oil refineries, cement-making works and chemical operations. All participants in the scheme are given a tradable emissions allowance (or credits) by the EU. One allowance gives the holder the right to emit one tonne of carbon dioxide or its equivalent.
- Polluters who find they have insufficient credits must either cut their emissions or incur fines, which will increase the price of their products. However, they may also buy unused credits from other participants. By setting a cap or limit on GHG omissions, the EU has created tradable allowances.
- This means that there is an incentive for businesses to decarbonize their operations because they will then be left with unused carbon allowances that can be sold for a potential profit.
- The scheme does not stop here, however. Over time, the number of allowances issued by the EU is reduced each year for all participants. This means that, over time, the entire economic system begins to decarbonize as more businesses adjust to life in a low-carbon economy.

The EU trading scheme has not been entirely successful, however. It has been criticized for issuing too many credits, which consequently have limited value, giving little incentive for participants to reduce their own use and sell surpluses. The high number of credits has also meant that little has been done overall to reduce the total flow of carbon emissions.

It is hoped that the EU's emissions trading scheme will become more widely used in the future and can deliver real reductions in GHG emissions ultimately. Carbon trading schemes are being adopted in other world regions too, as Figure 2.33 shows.

Figure 2.33 Carbon trading schemes

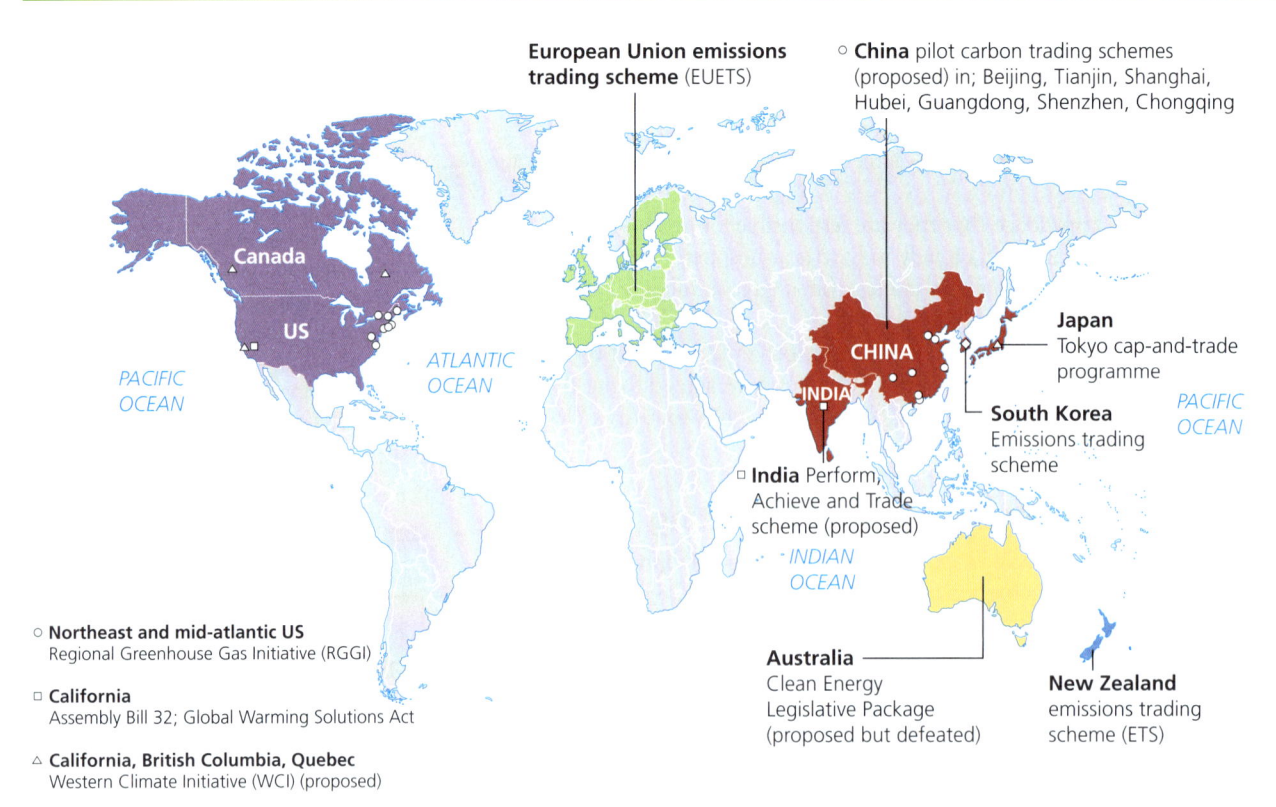

Map labels:
- European Union emissions trading scheme (EUETS)
- China pilot carbon trading schemes (proposed) in; Beijing, Tianjin, Shanghai, Hubei, Guangdong, Shenzhen, Chongqing
- Japan Tokyo cap-and-trade programme
- South Korea Emissions trading scheme
- India Perform, Achieve and Trade scheme (proposed)
- Australia Clean Energy Legislative Package (proposed but defeated)
- New Zealand emissions trading scheme (ETS)
- Northeast and mid-atlantic US Regional Greenhouse Gas Initiative (RGGI)
- California Assembly Bill 32; Global Warming Solutions Act
- California, British Columbia, Quebec Western Climate Initiative (WCI) (proposed)

Source: Stockholm Environment Institute; U.S. Energy Information Administration

Carbon offsetting

Carbon offsetting is another widely used mitigation strategy that aims to marry businesses principles with environmental goals. Our everyday actions – such as driving, flying and heating buildings – consume energy and produce carbon emissions. Carbon offsetting is a way of compensating for your emissions by funding an equivalent carbon dioxide saving elsewhere. The suggestion that we 'plant a tree' after taking an aeroplane flight is a well-known example of the offsetting principle.

- Various commercial organizations offer offsetting services: consumers pay a small, additional sum of money when they purchase a good or service; the company promises to invest this money in forestry, for instance.
- However, critics say there is insufficient monitoring of these schemes and the methods used to calculate how much carbon dioxide saving should be done. In the grand scheme of things, the carbon savings from offsetting do not amount to very much yet on a global scale either. Environmental writer Leo Hickman has argued that the whole concept of offsetting is 'a dangerous camouflaging of the real task at hand, namely reducing emissions'.

'Technological fixes' and geo-engineering

Technology is viewed by many people as the best way to 'fix' the problem of climate change. By asserting that yet-to-be-invented technology will solve the issues, it is easier to justify carrying on with 'business as usual'. In order to succeed, the Paris Agreement, as we have seen, relies heavily on technology that has yet to mature. In particular, it is hoped that technology can be used to remove part of the anthropogenic carbon dioxide store that is already in the atmosphere. This way, new anthropogenic flows of carbon dioxide can be accommodated for many years to come while developing, low-income countries use coal and oil to industrialize.

Table 2.8 provides an analysis and evaluation of several **geo-engineering** strategies that might offer a 'silver bullet' solution to climate change. Geo-engineering technologies aim to do one of two things:
- accelerate the removal of carbon dioxide from the atmosphere (carbon dioxide removal, or CDR)
- reflect more sunlight back into space (sunlight reflection methods, or SRM).

The *deliberate* intention to manipulate climate through these technologies distinguishes them from other human activities — such as coal burning or forest clearance — which *unintentionally* lead to changes in climate.

> **Keyword definition**
>
> **Geo-engineering** The deliberate, large-scale manipulation of the planetary environment in order to counteract anthropogenic climate change.

2.3 Responding to climate change and building resilience

Table 2.8 Possible 'technological fixes' for climate change using geo-engineering

Strategy	Analysis	Evaluation
Carbon capture and storage (CCS)	CCS involves capturing carbon dioxide released by the burning of fossil fuels and burying it deep underground (Figure 2.34). This technique promises to be extremely important (given that coal will remain a very significant part of the global energy budget for years to come due to its abundance and low cost). CCS works in three stages: 1. The carbon dioxide is separated from power station emissions. 2. The gas is compressed and transported by pipeline to storage areas. 3. It is injected into porous rocks deep underground or below the ocean for permanent storage (geo-sequestration). CCS could make an enormous difference to the size of the anthropogenic carbon store. The IPCC estimates that CCS (1) has the potential to reduce coal-fired power station emissions by up to 90 per cent and (2) could provide up to half of the world's total carbon mitigation until 2100.	• So far the technology has been piloted at only a handful of coal-fired power stations worldwide: it is far from being a mature technology. • CCS will be expensive because the technology is complex and still being developed. • There is uncertainty over how successful it will be. For carbon dioxide to remain trapped underground there must be no possibility of any leak to the surface. The gas cannot be a allowed to re-enter the atmosphere once it has been removed. • Pilot projects in the UK were cancelled recently owing to rising costs (of over US$1 billion). The plan had been for carbon to be transported by a pipeline to the North Sea and stored in depleted gas reservoirs. The UK government has cut public spending in many areas because of the global financial crisis and its after-effects.
Sunlight reflection methods (SRM)	SRM technologies aim to readjust the global energy balance. The hope is to reduce incoming solar radiation to offset the global heating caused by rising greenhouse gas emissions. These technologies include: 1. placing mirrors in near-Earth space orbit in order to reflect more sunlight back into space. 2. mimicking the global dimming effect (see page 39) of huge volcanic eruptions by injecting tiny sulfate aerosol particles into the stratosphere where they would scatter sunlight back to space 3. whitening low-level marine clouds by spraying seawater into them; the increased albedo (see page 40) would lead to more reflected sunlight.	• Costs and safety risks associated with the introduction of potentially millions of orbital mirrors are likely to be very high. • Stratospheric aerosol injections might disturb regional weather systems around the world, including storm systems. • Geopolitical tensions may arise if one country using this technology disrupts another country's weather. • Aerosol particles destroy stratospheric ozone (contributing to the 'ozone hole' effect). Ozone depletion allows cancer-causing ultraviolet radiation to penetrate the atmosphere. • Screening out some of the Sun's radiation will reduce the efficiency of solar power systems.

Source: Mike Hulme

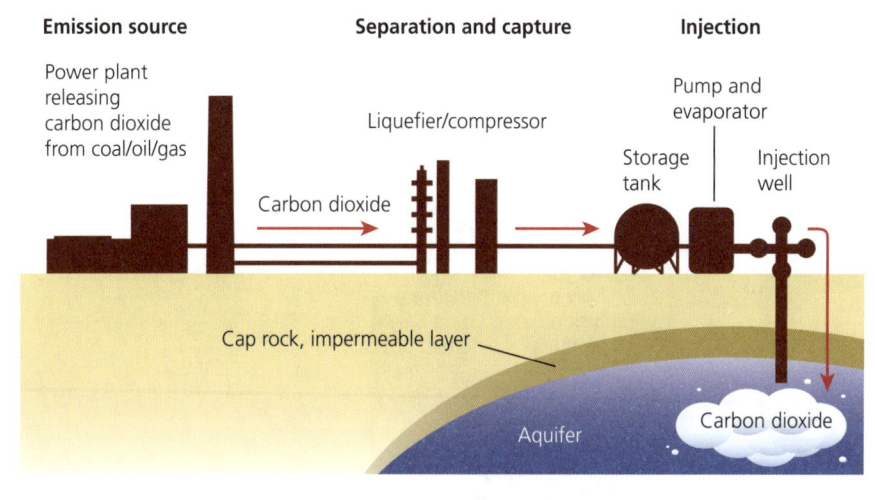

Figure 2.34 How carbon capture and storage (CCS) works

PPPPSS CONCEPTS

Think about how realistic are the technological possibilities shown in Table 2.8. Do you expect to see them implemented in your lifetime, and on what timescale?

Civil society and corporate strategies to tackle climate change

Revised

There is an urgent need for action at many different scales to deal with the threat of climate change. Governments have a crucial role to play in establishing frameworks within which low-carbon strategies can succeed. However, action needs to be taken additionally by individuals, civil society organizations (CSOs) and businesses alike now that we have arrived at a 'carbon crossroads' (Figure 2.35). Time is running out to adopt stronger mitigation measures.

Unfortunately, the complexity and long-term nature of the climate change threat means that some people, businesses and governments still do not view it as a priority for action.

- The global financial crisis of 2007–09 ushered in a new 'age of austerity' in high-income nations of the EU, Japan and the USA. With around one quarter of all young people unemployed in Ireland, Spain and Greece, many citizens are reportedly 'less concerned' with climate change than they were prior to the crisis. For many people in rich countries, paying their bills and housing costs has recently become more challenging. They may now be more reluctant to buy local food or replace their household appliances with 'greener' models if there is greater expense involved. As Table 2.8 showed, even the UK – one of the world's leading economies – has cut back spending on CCS research.
- People living in poverty in the world's 50 least developed countries – as well as around 0.8 billion Indians who live on less than US$2 a day – may still be far more concerned with day-to-day survival than they are with tackling a long-term challenge.
- Some people do not believe credible reports about climate change in the media and label it 'fake news' instead. The resulting tension in civil society may threaten our ability to respond to the challenges ahead effectively.

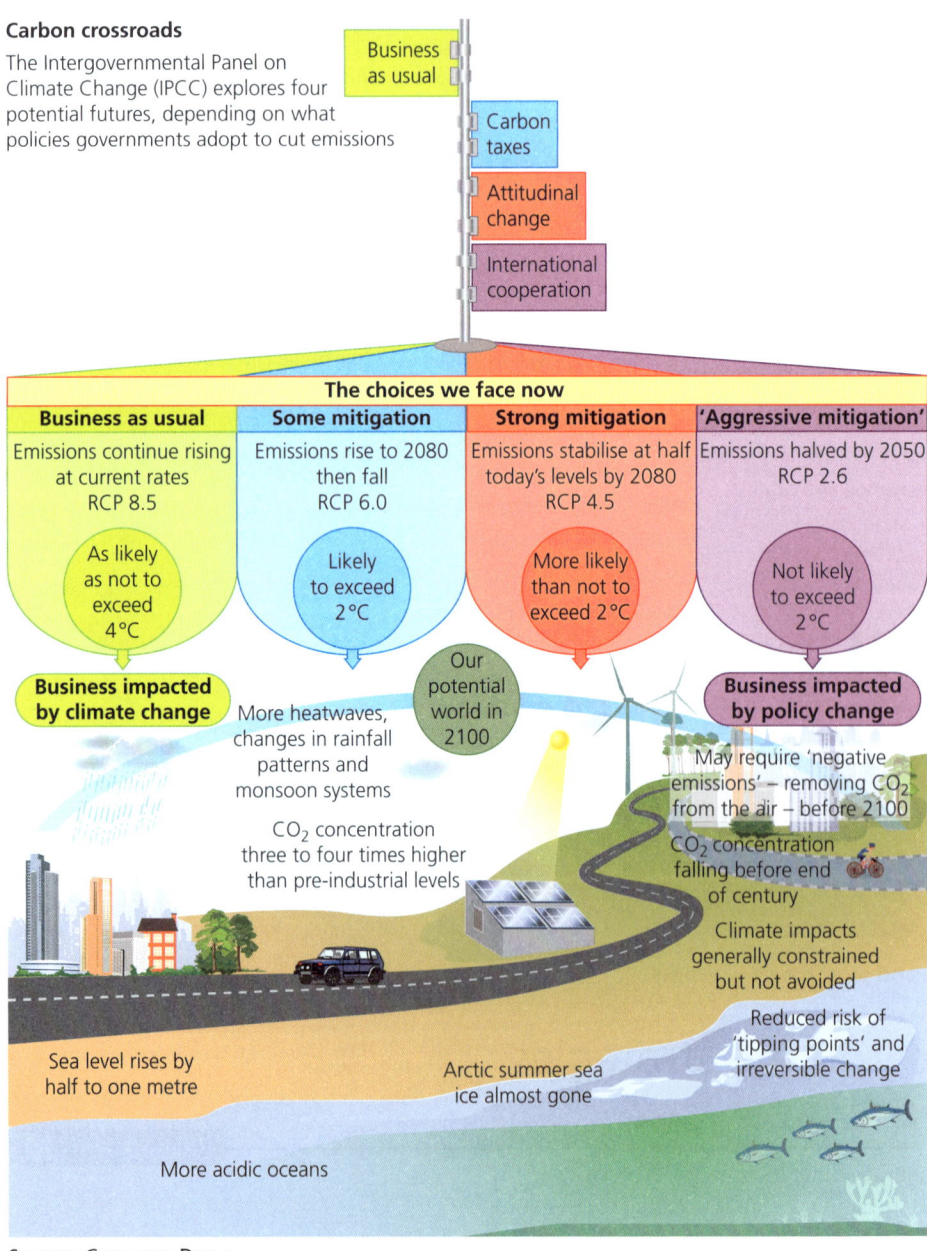

Carbon crossroads

The Intergovernmental Panel on Climate Change (IPCC) explores four potential futures, depending on what policies governments adopt to cut emissions

- Business as usual
- Carbon taxes
- Attitudinal change
- International cooperation

The choices we face now

Business as usual	Some mitigation	Strong mitigation	'Aggressive mitigation'
Emissions continue rising at current rates RCP 8.5	Emissions rise to 2080 then fall RCP 6.0	Emissions stabilise at half today's levels by 2080 RCP 4.5	Emissions halved by 2050 RCP 2.6
As likely as not to exceed 4°C	Likely to exceed 2°C	More likely than not to exceed 2°C	Not likely to exceed 2°C

Business impacted by climate change
- More heatwaves, changes in rainfall patterns and monsoon systems
- CO_2 concentration three to four times higher than pre-industrial levels
- Sea level rises by half to one metre
- Arctic summer sea ice almost gone
- More acidic oceans

Our potential world in 2100

Business impacted by policy change
- May require 'negative emissions' – removing CO_2 from the air – before 2100
- CO_2 concentration falling before end of century
- Climate impacts generally constrained but not avoided
- Reduced risk of 'tipping points' and irreversible change

Source: Cameron Dunn

Figure 2.35 The carbon crossroads (RCPs, or Representative Concentration Pathways, are different climate change projections used by the IPCC)

CASE STUDY

NON-GOVERNMENTAL AND CORPORATE ACTION IN THE USA

Many high-profile US politicians, including President Trump, are climate change skeptics. As a result, citizens who care about the issue believe it is more important than ever to act in whatever ways they can to reduce carbon emissions (no matter how small the scale of action). Table 2.9 shows a selection of non-governmental US stakeholders and analyses the actions they have taken in response to climate change. Despite the US government's 2017 decision to withdraw from the Paris Agreement, civil society and business stakeholders believe they can still make a difference and are acting accordingly.

Table 2.9 Examples of civil society campaigning in the USA

Individuals, businesses and places	Actions analysis
NextGen Climate (a civil society organization)	■ NextGen Climate in an environmental pressure group whose mission is to engage politically with Millennials about the connected issues of climate change and clean energy. They advise potential voters on which politicians share their environmental concerns. ■ The organization has a field operation in Las Vegas, which is viewed as being on the 'front line' of climate change in the USA: Las Vegas has doubled its consumption of water twice since 1985 and has suffered from severe droughts in recent years including 2016 (Figure 2.36).
Citizens	■ In recent years, increasing numbers of Las Vegas citizens and garden businesses have begun to adapt to what they perceive to be a permanent change towards even more arid conditions. Homeowners favour drought-tolerant 'desert landscaping' and are abandoning water-hungry grass lawns. ■ In colder states such as Montana, individual actions could involve turning down a home's thermostat by 1°C; this brings a 3 per cent reduction in total household energy use. This action cannot be forced through legislation and relies on action by individual citizens. Government can, however, play a role by educating people about the issues.
ExxonMobil (an oil and gas TNC)	■ Despite having much to lose if people were to abandon the use of fossil fuels, some US energy companies play an active role in supporting geo-engineering. Indeed, a corporate strategy of supporting CCS technology may be essential to their long-term profitability: it would mean consumers can keep using oil and gas while trusting CCS to remove anthropogenic carbon from the atmosphere. ■ ExxonMobil has a long record of investing in CCS research: http://corporate.exxonmobil.com/en/technology/carbon-capture-and-storage/carbon-capture-and-storage/developing-cutting-edge-technology-carbon-capture-and-storage
Seattle	■ The city of Seattle has its own Climate Action Plan (CAP), which aims to make the city carbon neutral by 2050. Adopted in 2013, Seattle CAP focuses on city actions that reduce greenhouse emissions and also 'support vibrant neighborhoods, economic prosperity and social equity'. ■ Actions are focused on areas of greatest need and impact – road transportation, building energy and waste: http://www.seattle.gov/environment/climate-change/climate-action-plan

PPPPSS CONCEPTS

Think about who has the greatest power to make a difference in relation to climate change. Is it the United Nations? The US President? TNCs? What arguments and evidence could help you make a case for each of these choices?

Figure 2.36 Las Vegas is on the 'front line' of climate change.

Unit 2 Global climate – vulnerability and resilience

> **■ KNOWLEDGE CHECKLIST:**
> - Social and geographic disparities in climate change risk and vulnerability
> - Examples of countries and communities with contrasting vulnerability to climate change, including the Philippines, the UK (London) and Greenland (Inuit communities)
> - The difference between adaptation and mitigation strategies used by governments
> - The global geopolitical response to climate change, including the Kyoto Protocol and Paris Agreement
> - Carbon emissions trading schemes (EUETS) and the value of offsetting
> - The importance of geo-engineering as a possible 'technological fix' for climate change, including carbon capture and storage (CCS) and sunlight reflection methods
> - Civil society and corporate strategies for climate change (using the USA as an example)
>
> **EVALUATION, SYNTHESIS AND SKILLS (ESK) SUMMARY:**
> - How perspectives on the need for climate change differ among individual and societies
> - How views vary on the possibility of whether dangerous environmental changes can be avoided
> - How optimistic scenarios rely on the power of technology to provide a 'fix' for climate change
> - How action can be taken at the local and not just the national scale.

EXAM FOCUS

SYNTHESIS AND EVALUATION WRITING SKILLS

Below is a sample answer to an exam-style essay question. Section C of your examination offers a choice of two essay questions for you to choose from. Having made your choice, you need to plan carefully an essay response which **synthesizes** and **evaluates** information (under assessment objective AO3). You can draw on ideas from any of the chapters in this book. This means that as you progress through the course, a greater volume of ideas, concepts, case studies and theories becomes available for you to synthesize. This sample answer draws on ideas from all of the first six chapters of this book.

The question has a maximum score of 10 marks. Read it and the comments around it. The levels-based mark scheme is shown on page vi.

'The world's poorer countries are least responsible for climate change and have most to lose because of it.' To what extent do you agree with this statement? [10 marks]

The amount of greenhouse gases in Earth's atmosphere has been increasing ever since the Industrial Revolution in Europe in the 1700s. Recently the concentration passed 400 ppm. If the global temperature rise is not kept below 2°C, the predicted impacts of sea level rise and temperature and rainfall changes could be catastrophic for countries everywhere and not just the poorest ones, as this essay will argue. (1)

The countries with the most to lose socially and economically include those whose physical position makes them extremely vulnerable to the impacts of climate change. This includes islands like the Maldives and low-lying countries like Bangladesh, much of which is very close to sea level. Temperature rises are predicted to be highest at high latitudes and this means there is high vulnerability too for Inuit communities in the Arctic Circle (as sea ice melts, they lose their hunting grounds). Finally, all countries with coastlines such as France, the UK, the USA, China, Nigeria and South Africa are vulnerable to sea-level rise. (2)

It is true that many countries with coastlines and high vulnerability to climate change are low- or middle-income countries. Historically, these nations have contributed less to the anthropogenic carbon store in Earth's atmosphere than high-income European and North American countries. The developed countries had their Industrial Revolution long ago. Factories were fuelled by coal burning and these countries must accept historic responsibility for much of the rise in atmospheric carbon dioxide over the last 250 years. Therefore we can accept that the statement is broadly correct, because countries like Tanzania, Jamaica and

1. This is a good, clear start, which introduces the concept of climate change and uses evidence expertly. The basis of an argument is also established.

2. This paragraph argues in favour of the statement convincingly. Each point is exemplified with a named country, which strengthens the case that is being made.

3. Here there is strong place-based knowledge of high-income nations and the historical period when they developed. There is a strong focus on the question that is being asked and constant reference is made to the idea of 'responsibility'. This shows that the answer should score well for AO2 (applied knowledge).

Vietnam – all of whom are threatened by sea level rises – have historically emitted relatively little greenhouse gas. (3)

It is also true that many low-income countries that are threatened by climate change have limited power to do anything about it. They have less political influence globally than the advanced high-income countries. Big TNCs are rarely headquartered in low-income countries and it is these companies that will also have a great deal of power over what will happen, including oil and gas companies. These global giants are headquartered in the USA, Russia, the UK and other developed countries. (4)

However, it Is also true that some of the highest emissions today come from middle-income countries. China is the world's largest polluter. India is increasing its emissions because more coal is being burned. Indonesia and Brazil are both responsible for higher rates of deforestation. This removes an important carbon sink from the climate system. What happens in these countries will very much determine whether or not GHG emissions cross the 550 ppm threshold, beyond which very severe impacts occur. Therefore they do have some responsibility and power over what will happen. (5)

Furthermore, some high-income countries have a great deal to lose because of climate change. Water shortages in US cities like Phoenix and Las Vegas are predicted to become severe. London's Thames flood barrier could fail if sea-level rises continue. Finally, the Inuit populations I mentioned earlier are often the citizens of high-income countries like the USA and Scandinavian nations. Poorer communities are found at the local scale in high-income countries, which complicates the argument further. (6)

In conclusion, the statement is only partially correct. While it is true that high-income countries started the problem of anthropogenic climate change, many poorer countries are now major polluters as well. Also, the effects of climate change could be highly damaging for countries everywhere and not just poorer ones. (7)

4 By shifting the focus away from states to the companies that are headquartered in those states, the answer shows a good grasp of the concept of power. Larger TNCs may have greater economic power than small, low-income countries, and it is useful to link them to the discussion. However, this paragraph does not feel as if it focused sufficiently on 'responsibility'. How could you modify this paragraph in order to improve its focus?

5 Here, a counter-argument has been advanced. The word 'however' has been used strategically to show the reader that a second point of view is now being put forward. This shows that the answer is well structured and deserves AO4 credit. There is also strong knowledge and understanding of emerging economies (they are 'poorer' countries than high-income European states, so this is a valid line of argument). A very good point is also made about the scale and degree of climate impacts that may occur. For this possibility to be avoided, all countries must take some responsibility.

6 'Furthermore' is another good word to use when constructing the counter-argument. Excellent points are made here, too, that invoke the geographic concept of scale.

7 This style of essay question requires a formal conclusion. Although short, the conclusion provided here does its job adequately.

Examiner's comment

Overall, this is a very strong essay that should reach the highest mark band. It is a balanced, structured and well-evidenced essay. Use is made of key geographic concepts. The writing is critical and reflective and thus meets the AO3 criteria.

Content mapping

You are now two-thirds of the way through the course and there is a large amount of material that can be drawn on to help answer this essay. This answer has made use of case studies, concepts, theories and issues drawn from all of the first six chapters of the book. Using the book's contents page, try to map the content.

Unit 3: Global resource consumption and security

3.1 Global trends in consumption

The concept of natural resources was introduced in Unit 1.1 in the context of the study of population distribution. In the past, geographers were concerned with the balance between (1) natural resources, (2) the number of people on the planet and (3) role of technology in 'unlocking' new resources (thereby allowing more people to live off the land). According to that view, the possibility of **sustainable development** (see Unit 3.3) was determined by the interrelationship between population size, natural resource availability and access to technology.

Now, it is widely recognized that a fourth element – **consumption** – must be added to this equation (Figure 3.1). This is because different individuals and societies consume resources in varying amounts and at contrasting rates. Changes in one population's lifestyle (as the country develops economically) can have a greater impact on global levels of resource use than population growth in a poor country might.

This chapter explores consumption patterns, trends and issues for three vitally important and interrelated categories of natural resource: water, food and energy (Table 3.1).

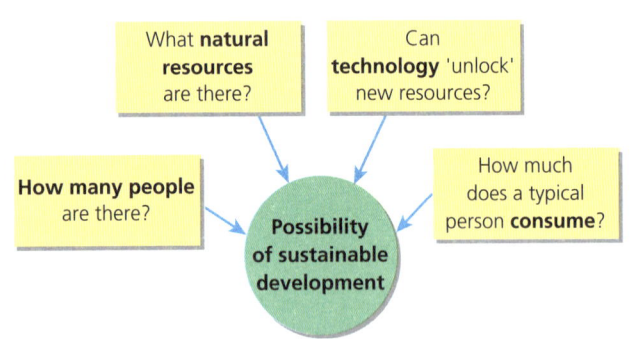

Figure 3.1 It is important to factor in consumption when we study the relationship between population, resources and sustainable development

Table 3.1 An overview of the importance of water, food and energy resources

Water	Food	Energy
Water is essential for human settlement and activity. Where water resources are scarce, people have learned to adapt to life with scarcity. However, a series of changes threaten local and global **water security** including: population growth; new lifestyle pressures that place greater demands per capita on water supplies; the appropriation of some people's water by other societies in an interconnected and globalized world; and climate change.	Without sufficient food intake, people suffer undernourishment and, in turn, undernutrition (which can lead to a person being underweight for their age and deficient in vitamins and minerals). About 800 million people are undernourished globally and the UN's Food and Agriculture Organization (FAO) is committed to tackling this. However, by 2050, food demand may double worldwide due to changing lifestyles and some population growth. **Food security** may become threatened.	Energy is vital for subsistence societies: fuel is required to keep people warm and cook food. The needs of industrialized societies are greater: additionally, people require energy to light their homes, power their transport and manufacture commodities they have become dependent upon. Experts predict a 40 per cent increase in global energy demand by 2040. To safeguard global **energy security**, over half a trillion dollars of investment is needed every year until then.

Keyword definitions

Sustainable development Meeting the needs of the present without compromising the ability of future generations to meet their own needs.

Consumption The level of use a society makes of the resources available to it. Economic development and changing lifestyles and aspirations usually result in accelerated consumption of resources.

Water security When all people, at all times, have sustainable access to adequate quantities of acceptable-quality water for sustaining livelihoods, wellbeing and development.

Food security When all people, at all times, have physical, social and economic access to sufficient, safe and nutritious food that meets their dietary needs and food preferences for an active and healthy life. Based on this definition, four food security dimensions can be identified: food availability, economic and physical access to food, food utilization and stability over time.

Energy security When all people, at all times, enjoy the uninterrupted availability of the energy they require to meet their needs, and at an affordable price.

Poverty reduction and the new global middle class (NGMC)

One of the most remarkable feats in human history has been the lifting of nearly one billion people out of absolute poverty during the past couple of decades. Much of the growth has taken place in Asia and Latin America. Over 500 million people have been lifted out of poverty in China alone.

The World Bank is the most important source of information about extreme poverty today and sets the International Poverty Line. This measure was revised in 2015: a person is currently considered to live in **extreme poverty** if he or she is living on less than US$1.90 per day. According to this measure, the proportion of people living in extreme poverty worldwide fell from 43 per cent in 1981 to 14 per cent in 2011 (Figure 3.2). In 2016, the number of people in extreme poverty was estimated to have fallen below 800 million by the World Bank.

This means that 90 per cent of the world's population lives above the extreme poverty line. A growing number belong to the **new global middle class (NGMC)**: this means they earn or spend more than US$3,650 a year, or US$10 a day. A further 2 billion people belong to a poorer group called the **fragile middle** class. They earn or spend between US$2 and US$10 a day: theirs is a precarious position and they might slide back into extreme poverty easily, were an economic crisis or major natural disaster to occur where they live.

Data source: World Bank

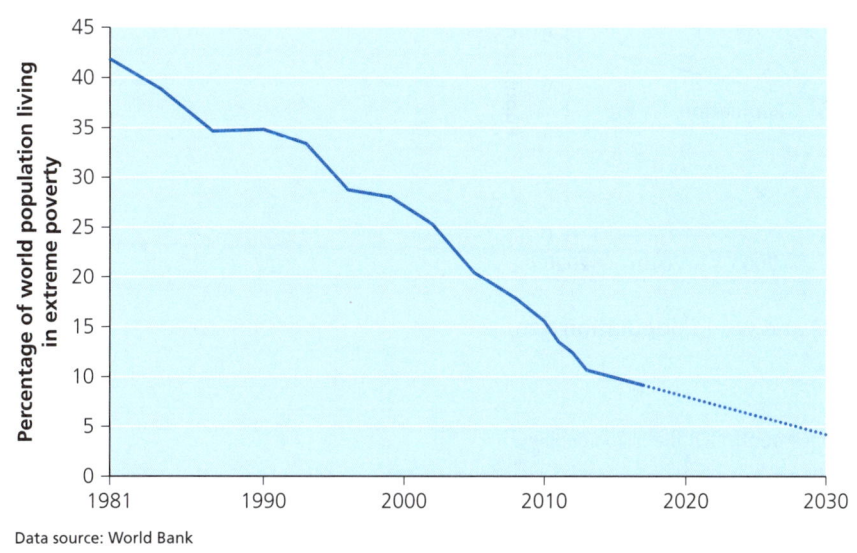

Figure 3.2 The falling share of the world's population living in extreme poverty (earning less than US$1.90 a day), 1981—2030 (projected)

> **Keyword definitions**
>
> **Extreme poverty** When a person's income is too low for basic human needs to be met, potentially resulting in hunger and homelessness.
>
> **New global middle class (NGMC)** Globally, the middle class is defined as people with discretionary income. They can spend this on consumer goods and, at the upper end, private healthcare, holidays or even cars. Precise definitions vary: one classification is people with an annual income of between US$3,650 and US$36,500; an alternative classification is people earning more than US$10,000 annually.
>
> **Fragile middle class** Globally, there are 2 billion people who have escaped poverty but have yet to join the so-called NGMC. This fragile middle class is broadly similar to the idea of a 'lower middle' class.
>
> **Millennium Development Goals** A set of interrelated global targets for poverty reduction and human development. They were introduced in 2000 at the UN Millennium Summit; their successor, the Sustainable Development Goals, followed in 2015.

■ Global poverty reduction

Overall, the global economy has grown enormously since the mid-twentieth century, and far faster than population (Figure 3.3). Global poverty has been halved since the introduction of the **Millennium Development Goals** in 2000, with the greatest progress made in Asia.

- About one billion fewer people lived in conditions of extreme poverty in 2017 than in 1990.
- The proportion of people in extreme poverty in developing countries (excluding China) fell from 40 per cent to 25 per cent between 1990 and 2010. When China is included, the figures are 4 per cent and 22 per cent – even more impressive. This shows that change there has played a key role when it comes to global poverty reduction targets being met.
- However, poverty remains widespread in sub-Saharan Africa and some parts of southern Asia. The poverty rate in sub-Saharan Africa fell only 8 percentage points over the same period.

The reasons for these changes are complex but relate in part to globalization and trade, in addition to work carried out by the United Nations and other international agencies. Richard Freeman of Harvard University has attributed some of the changes to a population dynamic he calls 'the great doubling'. The global labour force doubled to 3 billion people when China, India and Eastern European countries began to participate more fully in the world economy after a series of political changes in the 1980s. Another theory is that emerging economies have enjoyed the 'trickle-down' of scientific and medical know-how from Europe, North America and Japan.

In truth, there are many reasons for the broad 'headline' of global poverty reduction, and poverty trends in different countries should be analysed on a case-by-case basis. According to globalization critics, it may also be the case that the 'success story' of poverty reduction has been overstated. Figure 3.4 shows how data can be used selectively to present a number of contrasting viewpoints on poverty in the world today. Review the evidence and ask yourself: 'is the world becoming more or less unequal?'.

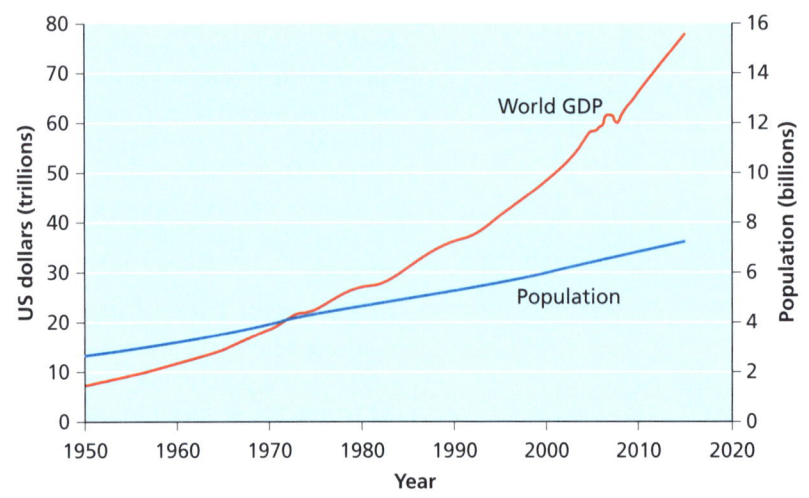

Figure 3.3 Comparing growth in world GDP and world population size, 1950–2015

Poverty in sub-Saharan Africa remains at nearly **50%** — no lower than in 1981; and population growth has meant the *number* of poor living here has doubled, from 200 million to about 400 million.	Poverty in East Asia fell from 80% of the population living below US$1.90 a day in 1981 to 18% in 2005. Much of the progress was in China, where **0.5 billion** people have been lifted out of extreme poverty.	Of the billion people worldwide who have escaped **US$1.90** a day absolute poverty since the 1980s, most would still be deemed as being very poor by European and North American standards.
The richest **eight** people alive today hold personal wealth equivalent to that possessed by the poorest half of humanity, numbering **3.6 billion** people in total.	The global GDP of **US$80 trillion** is shared unevenly between rich and poor countries. It is also shared unevenly among rich and poor people living within different countries.	**800 million** people live on less than US$1.90 a day (the World Bank extreme poverty measure); and **2 billion** people get by on just US$2–10 a day (purchasing power equivalent).
Around 350 million Indians lived in US$1.90 poverty in 2017; yet 41 of the 1,000 richest people on the planet were also Indian citizens – including **two** of the **top 50** earners.	The world's richest nations are also home to 100 million people who live below these places' official poverty line; including **45 million** US citizens (in 2014).	Between 1988 and 2011, the incomes of the poorest 10% increased by just US$65, while the incomes of the richest 1% grew by US$11,800 – **182 times as much**.

Figure 3.4 Analysing evidence of poverty and inequality at different scales

'The rich get richer while the poor get poorer'

Some critics of globalization claim that 'the rich get richer while the poor get poorer'. How can this assertion be reconciled with the World Bank's headline news about successful global poverty reduction? The answer lies in an analysis of *relative* differences in wealth. In 2017, Oxfam calculated that:
- the richest 1 per cent of the world's population has seen its share of global wealth increase from 44 per cent in 2009 to 99 per cent in 2016
- the richest eight billionaires possess the same wealth as the poorest half of humanity – that is 3.6 billion people! By this measure, the world has never been a less equal place.

Data like these suggest an 'explosion in inequality' at a time when nearly one billion people still live on less than US$1.90 per day. This indicates that development gap *extremities* (the range of values between the world's very richest and poorest people and countries) have increased.

The already-rich do take a disproportionately large share of each year's *new* economic growth. This is because they are in the best position to invest their existing capital in new opportunities, such as rising property prices in world cities like London or Beijing. These tripled in value between 2005 and 2015. In contrast, the majority of ordinary people's wealth and incomes have grown more slowly over time because they have fewer financial assets to invest in wealth-making purchases.

All of this means that as extreme poverty has fallen, the number of people living in **relative poverty** has risen in many societies. When the assets and earnings of the hyper-rich balloon in value, the *average* (per capita) level of wealth rises. As a result, some poorer people – whose earnings are static or have risen modestly – are reclassified as having below-average incomes despite the fact that they have experienced no material decline in wealth. To summarize, 'the rich get richer while the very poorest do not' is perhaps a more accurate view.

> **Keyword definition**
> **Relative poverty** When a person's income is too low to maintain the average standard of living in a particular society. Asset growth for very rich people can lead to more people being in relative poverty.

The new global middle class

Figure 3.5 shows the number of people alive in 1988 and 2011. As you can see, total world population grew significantly (by around 2 billion); sub-Saharan Africa experienced the largest proportional rise in numbers. The diagram also shows the changing distribution of wealth within China, India and other groups of countries:
- The modal (most common) income has increased in most countries and regions except sub-Saharan Africa.
- The number and proportion of people with a middle-class income of US$3,650 or greater has risen.

Investigating the NGMC

Historically, the phrase 'middle class' came to describe a socio-economic group sandwiched between the workers (or 'working class') and the 'ruling' class in European countries. Today, the idea of a 'global middle class' is used to describe a growing mass of people who no longer experience the absolute poverty still endured by almost one billion people globally. However, they have not yet achieved the affluent lifestyles of 'the Western world' or Japan and South Korea.

Opinions vary on what exactly defines this 'middle class'. Essentially it describes those who are left with disposable income after essentials (shelter, heating, food) have all been paid for. At the bottom end of the middle-class income range, this might include someone who can afford to buy a non-essential can of Coca-Cola. At the upper end of the income range, it means having enough money to buy a fridge, smartphone or even a cheap car.

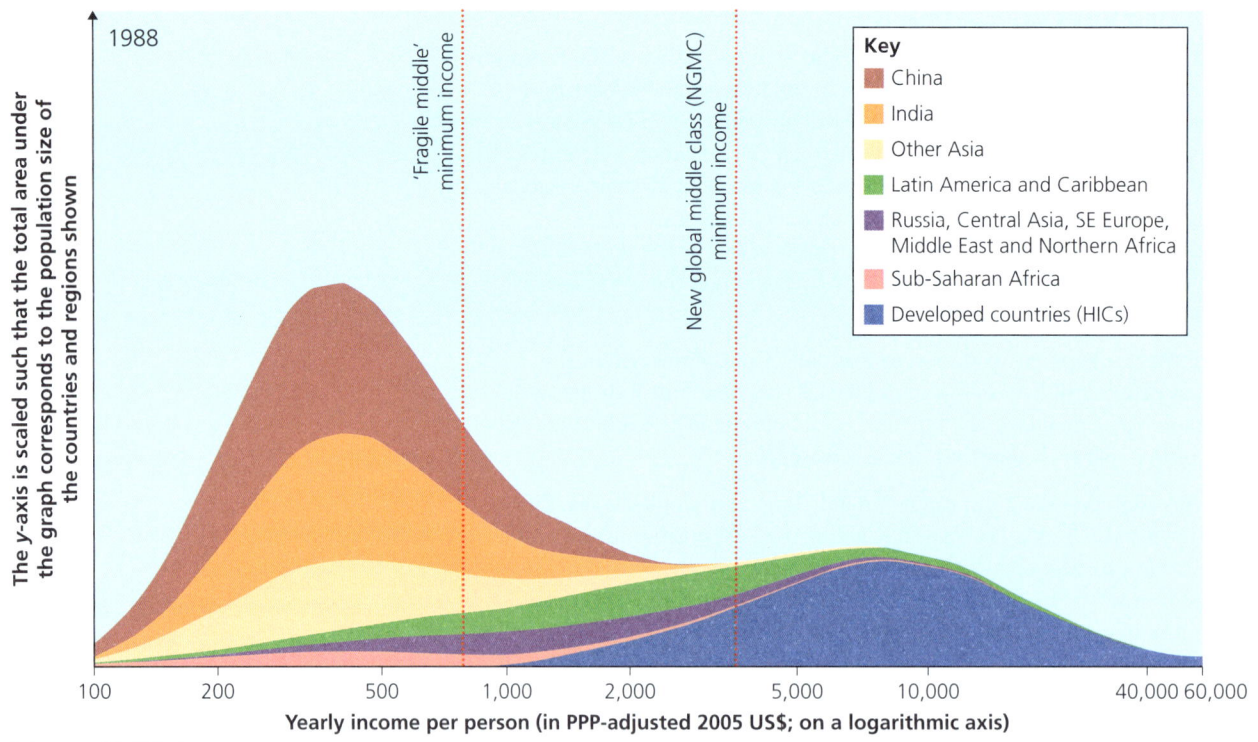

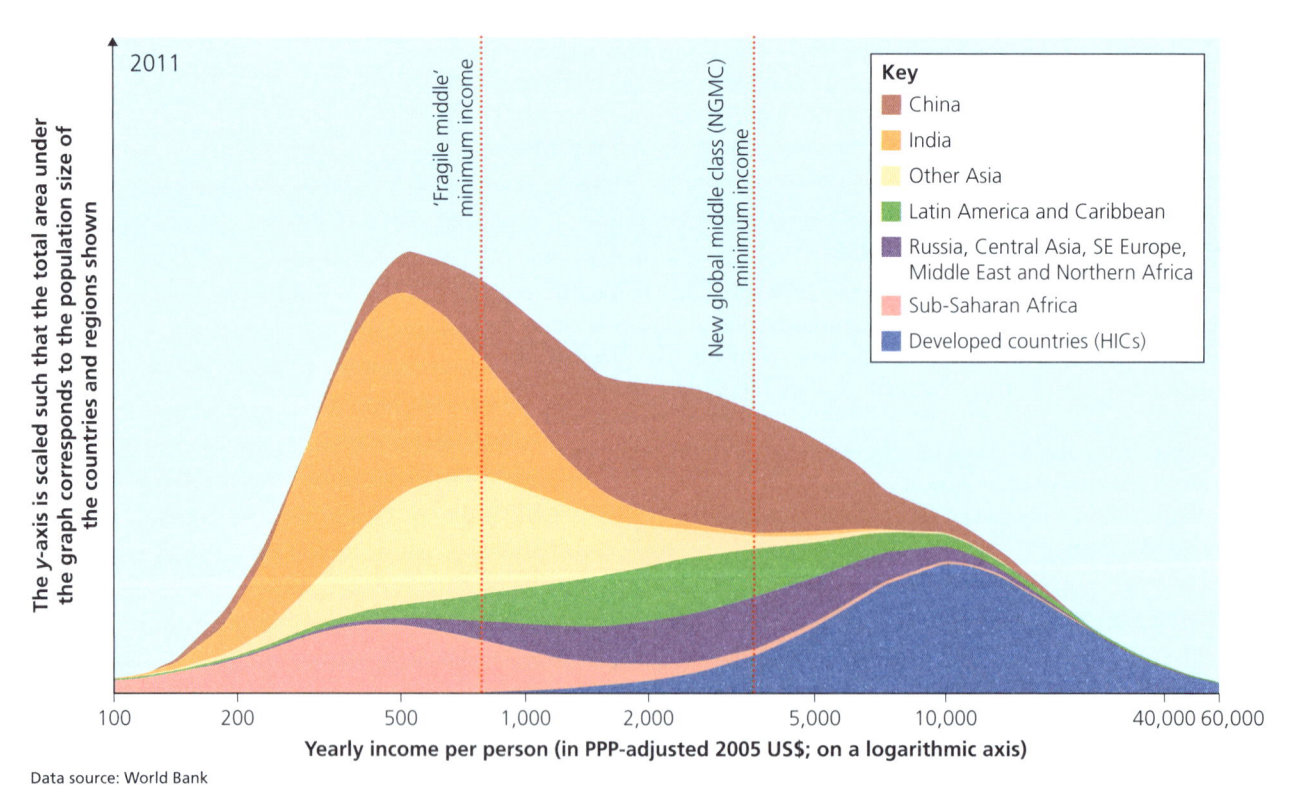

Figure 3.5 Changes in global income distribution 1988–2011 suggest a rising modal income in many places and a growing proportion of middle-class earners

- By one estimate there are currently 2 billion people in the NGMC (Figure 3.6), but this will grow to 3.2 billion by 2020 as more of the fragile middle class see their incomes rise.
- Asia is almost entirely responsible for this growth: its middle class is forecast to triple in size to 1.7 billion by 2020. By 2030, Asia will be the home of 3 billion middle-class people (10 times more than North America).
- There is also substantial middle-class growth in the rest of the emerging world (Table 3.2). The middle class in Latin America is expected to grow from 180 million to 310 million by 2030, led by Brazil. In Africa and the Middle East, it is projected to more than double, from 140 million to 340 million.

Remember, however, that estimating a country or population's income or spending is not easy (see Unit 1.1, page 5, for an account of the difficulties of calculating GDP data accurately). Different countries use different currencies. Also, money goes further in some places than others. Income figures for different countries are therefore adjusted – sometimes crudely – to take into account purchasing power parity (PPP). For example, the average income in China is roughly US$7,000 but this becomes US$10,000 when adjusted for PPP. It is therefore important to remember that some data used to report the NGMC phenomenon may not be entirely accurate or trustworthy.

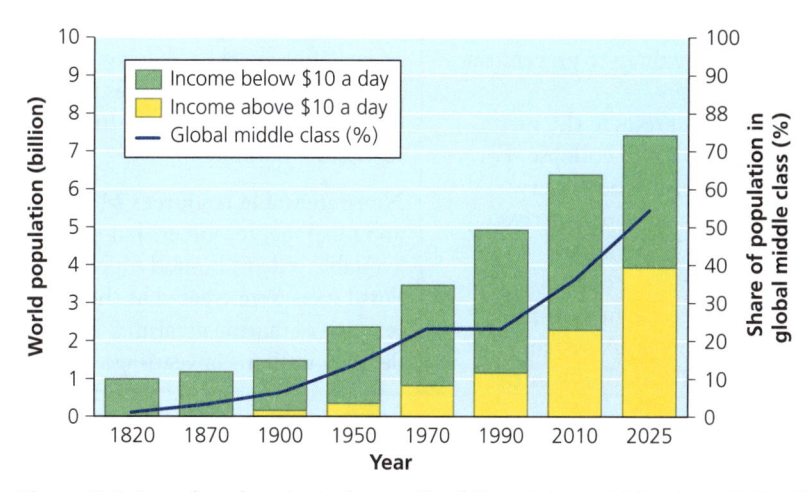

Figure 3.6 Actual and projected growth of the global middle class, 1820–2025

Table 3.2 Emerging economies and their growing middle class, 2014

Country	Population (millions)	Selected data
Indonesia	238	The middle class of people earning more than US$10 a day is predicted to grow from 45 million today to 135 million by 2030.
Mexico	112	65 per cent are middle class, each spending on average US$9,000 annually (Mexico is the USA's second-largest export market). Average wage was US$2.11 an hour in 2014.
India	1,210	Fewer than one in 20 people are middle class, but it could be 1 in 5 by 2025. India's retail market is worth around US$700 billion annually.
China	1,347	One in five are middle class and they spent US$1.5 trillion in 2012, making China the world's largest market for cars and mobile phones.
Brazil	194	The size of the middle class grew from 31 per cent of the population in 2000 to 47 per cent in 2012. Many are high earners of more than US$20,000 a year.
Total	3.1 billion	

Trends in resource consumption

Poverty reduction and the growth of the NGMC have led to a reappraisal of (1) what a 'typical' **ecological footprint** will look like by mid-century, and (2) the implications for future water, food and energy security. The present global average consumption of plant energy (as a direct food source, or used indirectly for animal feed) is 6,000 calories daily. On this basis, Earth's **carrying capacity**, by some estimates, stands at 11 billion people. But if average consumption rises to, say, 9,000 calories then the population carrying capacity of Earth falls to just 7.5 billion.

■ Renewable and non-renewable resource consumption

Some natural resources are **renewable**, including sustainably managed cropland and forest, wind power and solar energy. Others are **non-renewable**, including minerals and fossil fuels. The rate at which both renewable and non-renewable resources are used can have implications for future generations.

- Most water supplies and the biomass that provides food are renewable, provided safeguarding principles are applied to their management. For instance, hunting and fishing quotas need to be used sometimes to prevent the permanent loss of certain animal and fish populations.
- Over-fishing can leave insufficient fish in the sea to reproduce the next generation. The worst-case scenario is a complete fish stock collapse. For example, the collapse of the Newfoundland Grand Banks cod fishery in Canada in 1992 led to the loss of one of the world's most productive fishing grounds. The fishing industry, government and consumers collectively managed the ecosystem unsustainably. They increased their catch of fish (a system output) far more quickly than the natural replacement rate of young fish being born (a system input). A system threshold was eventually crossed, which led to the collapse of the entire fish stock. It has never recovered.
- Water sources like aquifers and inland lakes can be mismanaged similarly. The inland Aral Sea has all but vanished. More than any other water body in the world, it has come to epitomize the devastating economic and ecological effects of excessive demands placed upon freshwater water stores.
- Non-renewable resources, including minerals and fossil fuels, exist only in finite amounts. The best that can be hoped for is that (1) we can slow down the flow from these stocks, or (2) a technological 'fix' will allow new reserves to be discovered or recovered.

Population growth and rising affluence combine to make it harder than ever to conserve the use of renewable resources (thereby avoiding a 'tipping-point' from being reached, beyond which those resources may no longer be available). Population and development growth processes also make it more likely that the moment of **peak oil** will arrive soon. However, another point of view is that rising demand for resources – often accompanied by rising prices for commodities – stimulates research for alterative supplies and sources of water, food and energy. This theme is returned to in Unit 3.3.

> **Keyword definitions**
> **Peak oil** The point when the maximum rate of global oil production is reached (likely to be before 2030).

> **Keyword definitions**
> **Ecological footprint** A crude measurement of the area of land or water required to provide a person (or society) with the energy, food and resources needed to live, and to also absorb waste (Figure 3.7).
>
> **Carrying capacity** The maximum number of people an area of land can support with current levels of technology.
>
> **Renewable resources** Natural resources that are replenished by the environment over relatively short periods of time. Forestry is a renewable resource: regrowth occurs after wood has been cut provided the rate of use does not exceed the forest's capacity to regenerate naturally.
>
> **Non-renewable resources** Mineral and fossil fuel resources that are available only in limited supplies. Fossil fuels were created as the remains of marine creatures decayed millions of years ago, under huge amounts of pressure and heat. They cannot be replaced easily.

ECOLOGICAL FOOTPRINT ANALYSIS

Researchers look at how much land, sea and other natural resources are used to produce what people consume. They also consider how much land is needed to dispose of an individual or society's waste. The Global Footprint Network (GFN) organization has been attempting the tricky business of measuring ecological footprints since 2003 using the framework shown in this illustration.

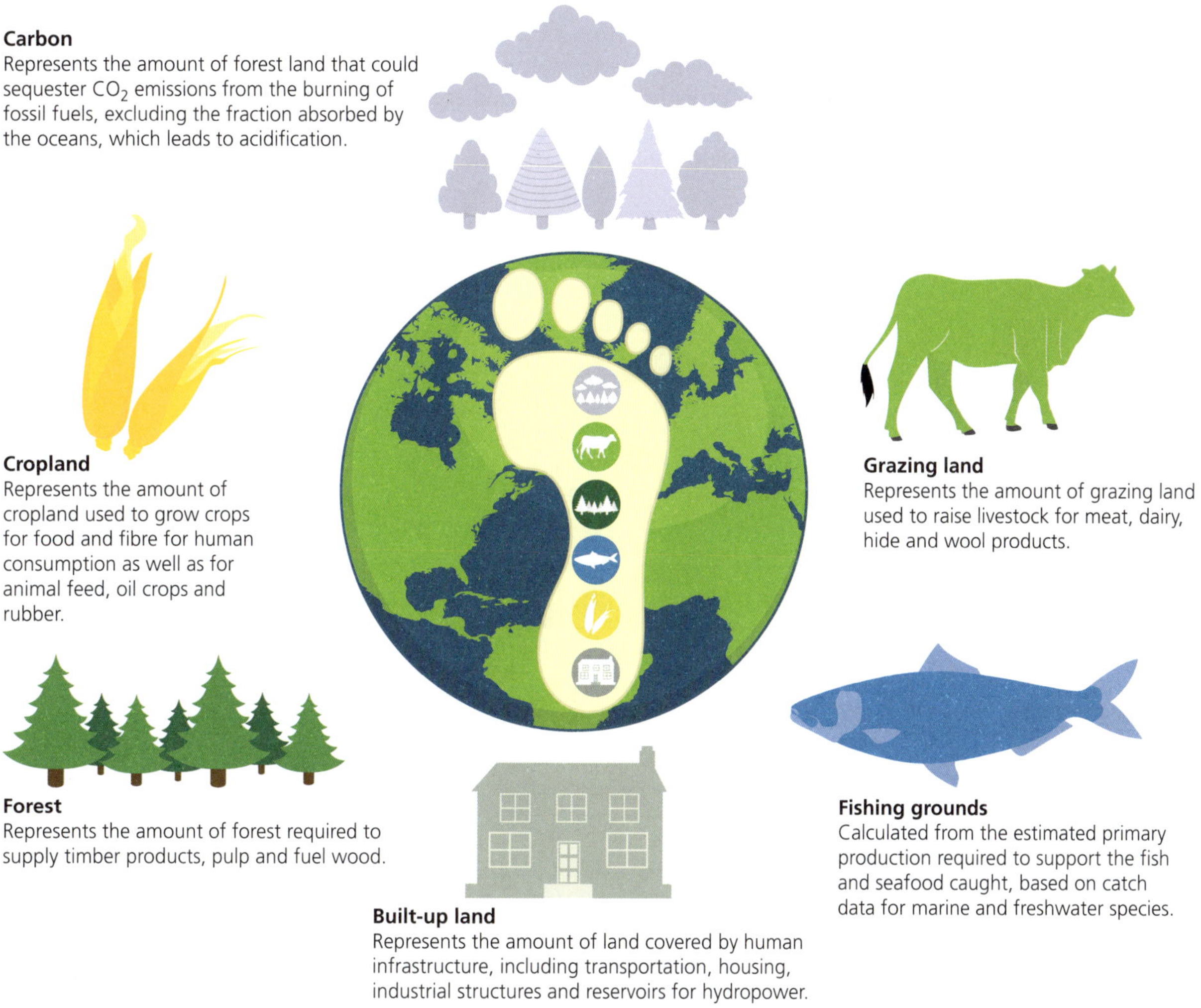

Carbon
Represents the amount of forest land that could sequester CO_2 emissions from the burning of fossil fuels, excluding the fraction absorbed by the oceans, which leads to acidification.

Cropland
Represents the amount of cropland used to grow crops for food and fibre for human consumption as well as for animal feed, oil crops and rubber.

Forest
Represents the amount of forest required to supply timber products, pulp and fuel wood.

Grazing land
Represents the amount of grazing land used to raise livestock for meat, dairy, hide and wool products.

Fishing grounds
Calculated from the estimated primary production required to support the fish and seafood caught, based on catch data for marine and freshwater species.

Built-up land
Represents the amount of land covered by human infrastructure, including transportation, housing, industrial structures and reservoirs for hydropower.

Figure 3.7 The ecological footprint concept

Ecological footprints are studied and calculated at the individual, national and global scale:

Individual	• Measured by energy consumption, today's average individual US citizen has a *per capita* ecological footprint 20 times larger than a person living in the pre-industrial 1700s.
	• In other words, the same area of land that supports 10 US individuals with high-impact lifestyles today could support 200 low-impact lifestyles.
	• If world population stabilizes at 9 billion around 2050 – and economic development eventually grants the majority of these people an 'American' living standard – this becomes equivalent to 180 billion 'pre-industrial' individuals, in terms of energy consumption.
National	• China is a state whose ecological footprint has expanded in size greatly.
	• Since China opened itself up to foreign investment in 1978, GDP has risen by a staggering 2,000 per cent. Rising wages have fuelled consumption of branded goods in shiny new shopping malls.
	• Since 1970, China's total ecological footprint has doubled in size and is now the largest in the world (the *per capita*, or individual, ecological footprint of Chinese citizens is half that of US citizens, however).
Global	• Since around 1970, the world has been in a state of 'ecological overshoot'. Earth's total biocapacity is 12 billion hectares (1.8 hectares per person) while humanity's ecological footprint is 18.2 billion hectares (2.7 hectares per person).
	• In other words, we are using the equivalent of 1.5 Earths to support our consumption. If we maintain our current lifestyle and consumption patterns, by 2030 we will need more than two Earths.

Patterns and trends in global water consumption

Water is the most fundamental human need; as we have seen, only 3 per cent of the world's water is non-saline and much of this is locked in ice (see page 49). With world population increasing in size and wealth, water consumption is rising. Growing demand for water comes from a mixture of urban consumers, agricultural producers, industry and hydropower dams (Table 3.3). Figure 3.8 shows that total freshwater withdrawals are predicted to increase by 2025 in all regions of the world.

- In 1900, total freshwater use was just under 600 cubic kilometres; by 1950 it had more than doubled.
- Total demand for water is projected to exceed 5,000 cubic kilometres by 2025.

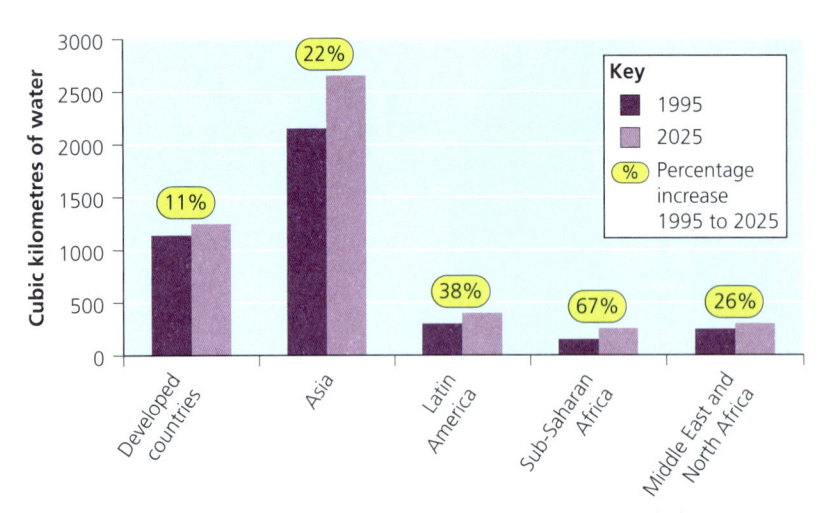

Figure 3.8 Patterns and trends of freshwater withdrawals by region, 1995 and 2025 (projected)

Table 3.3 Causes and symptoms of increased global water consumption

Agriculture, food and drink production	• Crop cultivation, crop processing, food distribution (and even the final recycling phase for food and drink packaging) all require water. • Agriculturally driven water stress is especially evident in the Indus river basin, which is home to the world's largest irrigation system. Irrigation accounts for over 90 per cent of water withdrawn from available sources for use in many semi-arid regions of the developing world. • At the Ogallala aquifer, which stretches from Texas to South Dakota, the water level is dropping by 90–150 cm per year. This decrease will threaten one third of irrigated agriculture in the USA within the next 40–180 years, with potentially huge impacts on grain supplies and prices.
Industry	• After agriculture, industry is the second largest user of water. • Emerging economy growth since the 1980s has increased the amount of industrial activity worldwide. According to World Trade Organization statistics, the value of world trade in industrial products rose from US$2 trillion in 1980 to US$18 trillion in 2011, which gives some idea of the extra pressure placed on water supplies in recent decades.
Household water consumption	• Poverty reduction and NGMC growth increases water consumption globally: more people than ever before expect to have unlimited access to safe water for drinking, bathing, showering and domestic appliances (washing machines); sanitation and sewerage systems also place a heavy demand on water supplies. • Worldwide, 2.6 billion people lack sanitation (those in poverty and most of the 'fragile middle'). However, as poverty alleviation increases further, more of this 2.6 billion are expected to make the transition to greater affluence (with the expectation of clean fresh tap water 24 hours a day, 7 days a week). This inevitably means more water consumption in many parts of the world where there are already water shortages, including North Africa and the Middle East.

Consumption of embedded water

The term **embedded water** describes the volume of water required to produce a crop, product or service (and which then becomes part of its 'water footprint'). For example:

- A can of cola contains 0.35 litres of water, yet it requires an average of 200 litres to grow and process the sugar contained in that can.
- It takes 2,900 litres to 'grow' a cotton shirt and 8,000 litres to produce a pair of leather shoes; that is, the amount of water required to grow feed, support and process that fraction of a cow that makes just one pair of leather shoes.

As a result of embedded water, the average European actually requires an amazing 4,600 litres of water to sustain his or her lifestyle on a daily basis. That number exceeds greatly the number of litres needed for bathing, hygiene, drinking and cooking. This is because the calculation takes into account all of the water used to grow crops and raise animals over their lifetime to provide the meat and dairy products many people eat each day. Factor in the additional water used by power stations and industrial processes and you will begin to see how our use of embedded water greatly enlarges the size of each person's individual 'water footprint'.

Figure 3.9 shows the global network of interconnected places that supply the UK with goods and food which have a high embedded water value. As you can see, the water demands of just one country turn out to be vast.

> **Keyword definition**
>
> **Embedded water** A measure of the amount of water used in the production and transport to market of food and commodities (also known as the amount of 'virtual water' or 'water footprint' attached to a product). Embedded water may include the use of local water resources and the use of water resources in distant places.

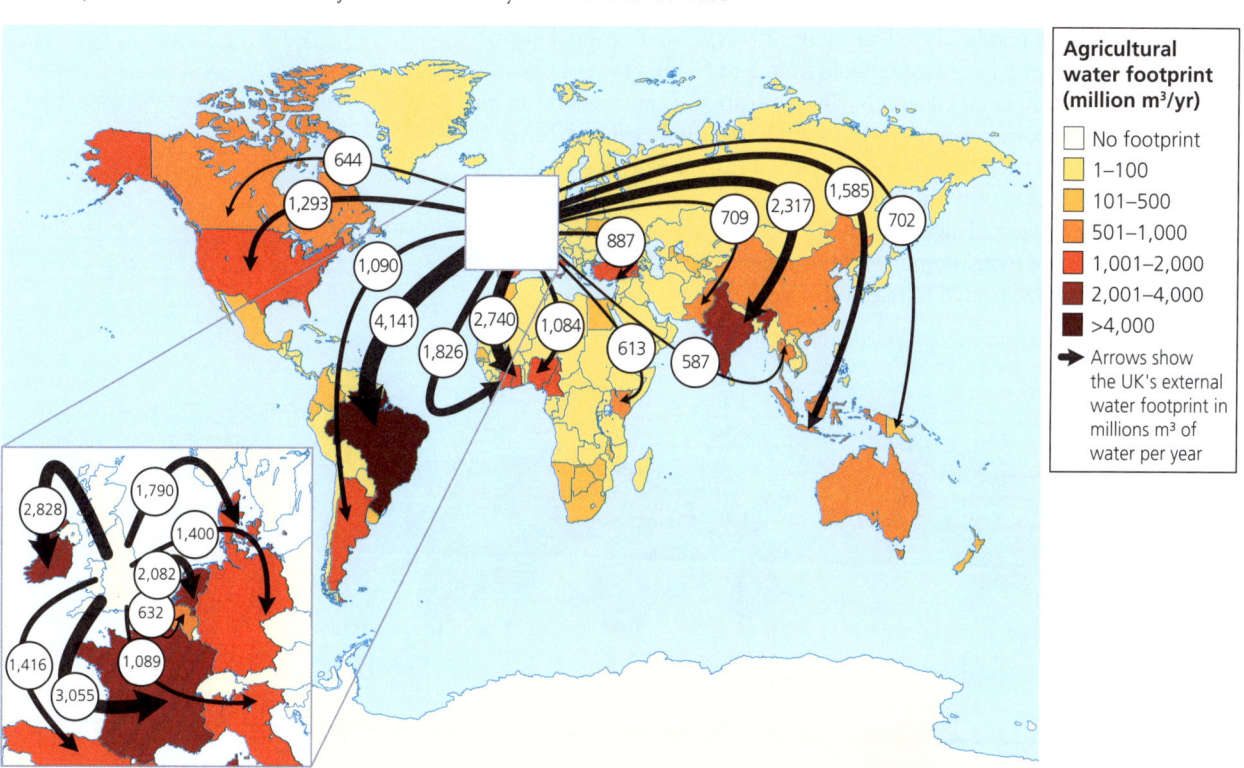

Figure 3.9 Embedded water in the goods and food the UK imports from other places

Different diets have very different water demands; as a result, people in high-income countries consume much greater volumes of embedded water than individuals in low-income countries. A meat and dairy diet uses many more litres of embedded water per day than a vegetarian diet. This is because crops that could be directly consumed by humans are eaten by animals which themselves require additional water to meet their hydration and respiration needs (the interrelations between food and water security are returned to in the next chapter). 'Luxury' drinks also have a high embedded water value. Table 3.4 shows the amount of water required to grow crops and complete the manufacturing process for some common beverages.

> **PPPPSS CONCEPTS**
>
> Think about the importance of the key geographical concept of spatial interactions for the study of global water consumption. Many places in the world rely on water from other places at varying scales. Some parts of a country draw their drinking supplies from another part of that country. They also eat imported food and goods from other countries, which have high levels of embedded water.

Table 3.4 Water needed to produce one litre of beverage (litres)

Coffee	Wine	Apple juice	Orange juice	Beer	Tea	Bottled water
1,100	960	950	840	300	130	5

Patterns and trends in global food consumption

With world population size growing by around 80 million a year, there are always more mouths to feed. By mid-century, there will be 9 billion people: far more food will be required than today. The productivity problem is compounded further by the **nutrition transition** to a meat and dairy diet which is taking place among the NGMC in emerging economies, along with a continued rise in per capita consumption of meat in high-income countries.

Figure 3.10 shows the global pattern of food consumption measured by daily calorie consumption per person.

- The relatively low values in much of Africa and parts of Asia, including India, reflect the fact that many people living in these regions still rely on a basic vegetarian carbohydrate diet (such as rice).
- The higher values seen elsewhere show how traditional diets (often low in meat and high in vegetables) are giving way to more meat and 'fast food' among the emerging middle class. During the 1990s, China's annual meat consumption per capita increased ten-fold from 5 to 50 kilograms. In Brazil, the figure rose from 30 to 80 kilograms. The populations of both countries have also become enthusiastic consumers of a range of imported food and drinks, including European cheese and Scotch whisky (and, as a result of such dietary changes, diabetes and obesity are on the rise in China and South America).
- Figure 3.11 shows projected growth in middle-class spending worldwide by 2030: this gives some indication of world regions where meat, dairy and luxury food consumption is expected to rise even further.

> **Keyword definition**
>
> **Nutrition transition** A change in diet from staple carbohydrates towards meat and fish proteins and dairy products. This happens typically when incomes rise from US$2 a day to US$10 a day.

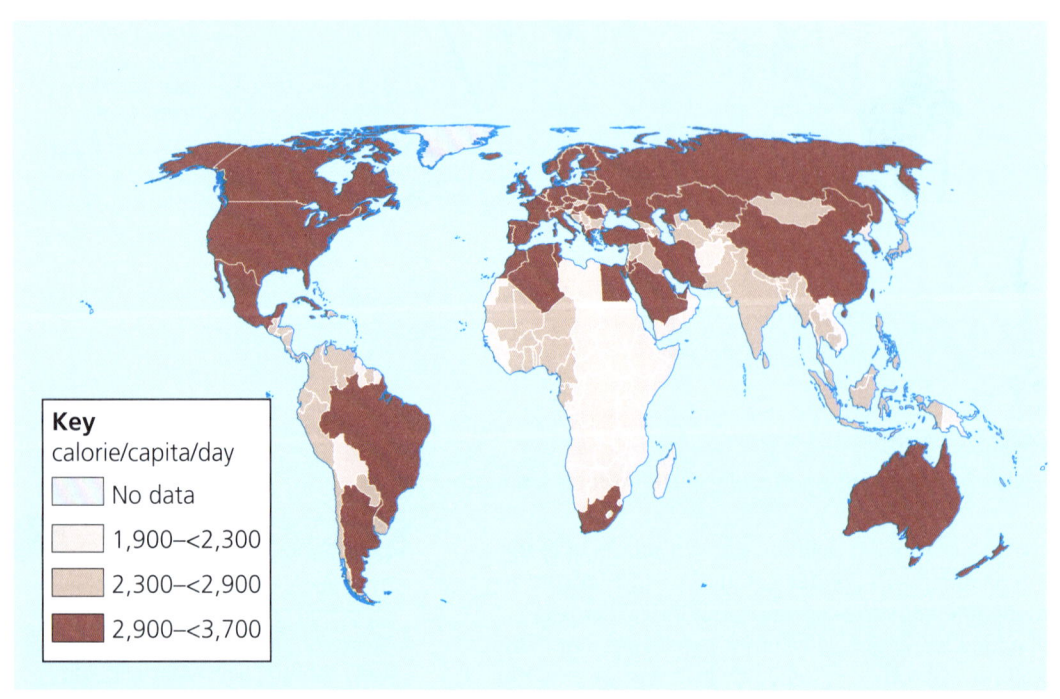

Figure 3.10 The global distribution of daily calorie intake per person, 2009–11 (FAO)

Middle class consumer spending
Outer ring: 2030 in trillions, USD (projected)
Inner ring: 2009 in trillions, USD

$11.1

$5.6

Europe
$8.1

Asia Pacific
$32.9

North America
$5.5

$4.9

Sub-Saharan Africa
$0.6

$0.4

$1.5

Middle East and
North Africa

$0.9

$3.3

$2.2

Central/
South America

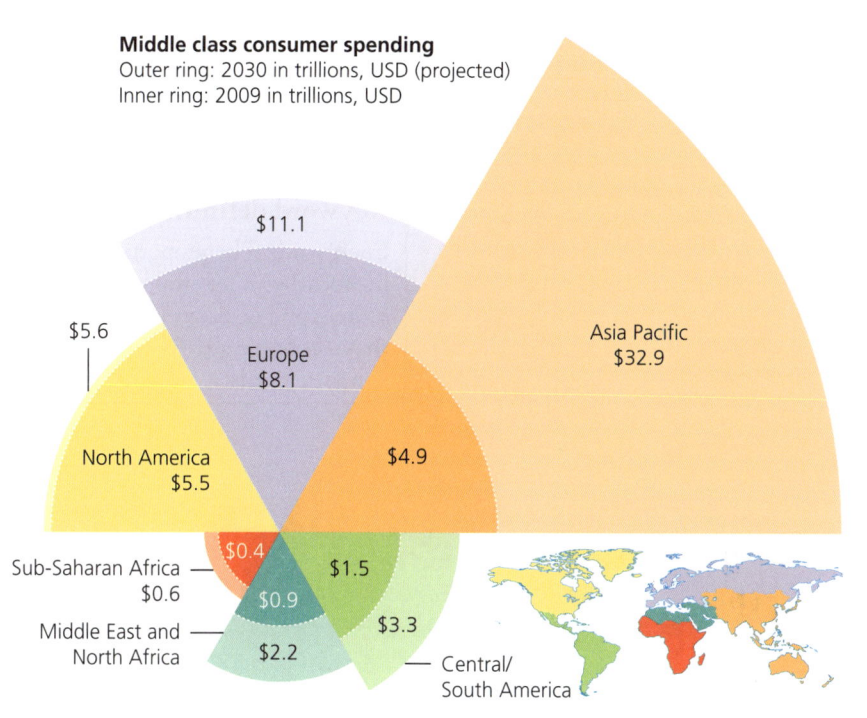

Source: Cameron Dunn

Figure 3.11 The growth of global middle-class spending by 2030 (with implications for food consumption)

Geographic implications of the NGMC's nutrient transition

The global pattern of daily calorie intake is only part of the picture of the increased pressure that rising affluence places on food production systems. The pressure is intensified even further by the inefficient way food chains operate:

- Animals use up a lot of converted biomass energy roaming around, defecating and respiring.
- Beef cattle eat about 8 kilograms of grain or meal for every kilogram of flesh they produce.
- Cattle rearing therefore places excessive and wasteful demands on grain (and water) supplies that could be used to provide for human populations directly.
- Fewer grains and cereals are left to be sold as food on global markets if large volumes are being used as inputs for wasteful cattle farming systems.

Dietary and lifestyle changes have further implications for agricultural production systems and the species they rely on:

- Livestock farming has become the new focus of Asian agriculture, bringing with it a steep rise in emissions of methane, a powerful greenhouse gas. This contributes to climate change, which may affect crop yields negatively in some parts of the world (see pages 58–59).
- Rising affluence also puts pressure on particular plant and animal species if their use or consumption is linked culturally with social prestige. Southeast Asia is being stripped of Siamese rosewood as a result of the ability of China's emerging middle class to purchase sought-after hardwood furniture that used to be out of their price range. Shark fin soup is an important but expensive dish traditionally consumed at Chinese weddings by those who can afford it. As incomes have risen, the number of sharks killed worldwide to meet growing demand has doubled. Unless action is taken, stocks of these and other renewable plant and animal resources may eventually collapse altogether.

PPPPSS CONCEPTS

Think about the varied ways in which animal farming has impacts on physical systems and processes at different scales.

Patterns and trends in global energy consumption

Every country meets its growing energy needs in a particular way, known as its **energy mix**. Figure 3.12 shows global trends in energy use. Total consumption has soared upwards, especially since the 1940s; you can also observe the individual trends for different types of energy. Total energy consumption can be measured in (1) kilograms or tonnes of oil equivalent per year, (2) gigajoules (or exajoules) per year, or (3) megawatt hours per year.

> **Keyword definition**
>
> **Energy mix** The proportions of hydrocarbons, renewable energy sources and nuclear energy that a country uses to meet its domestic needs.

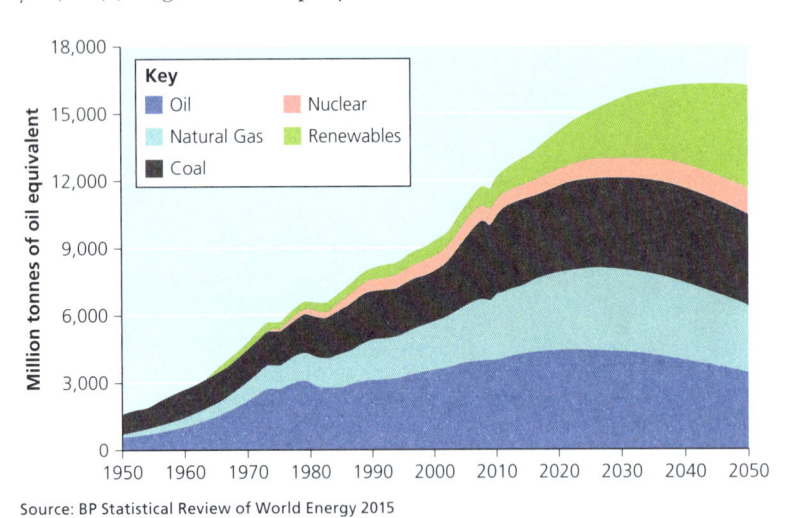

Source: BP Statistical Review of World Energy 2015

Figure 3.12 Global energy use 1950–2050 (projected)

Total energy demand is predicted to grow by around 40 per cent between 2000 and 2030 (Figure 3.12). This is on account of the continued rising energy use of the NGMC, and projected energy demands for the 2 billion fragile middle class who are poised to join the NGMC by becoming 'powered up' before 2030.

- Nearly 3 billion people worldwide still use wood for fuel, filling their homes with life-shortening smoke.
- Two-thirds of them could begin making use of electricity supplies in the near future, ending their reliance on traditional biofuel.

Energy mix patterns

Three main types of energy are used to produce electricity and generate heat. They are (1) **hydrocarbons**, (2) nuclear power and (3) **renewable** sources of modern energy. Table 3.5 provides a brief profile for each of these.

Table 3.5 Profiles of the three main types of energy source used globally

Hydrocarbons (non-renewable fossil fuels)	Hydrocarbon fuels contain carbon and hydrogen. During combustion, hydrogen is oxidized to produce water and carbon dioxide. Hydrocarbon fuels include: • 'conventional' oil and gas supplies found underground • coal (a combustible sedimentary rock, and the 'dirtiest' of all hydrocarbons in terms of carbon emissions) • 'unconventional' oil and gas supplies including shale gas, which is found trapped in sedimentary rocks thousands of metres underground. Shale gas is released by injecting a high-pressure mixture of water, sand and chemicals into the rocks. This process is called hydraulic fracturing – or 'fracking' (Figure 3.13).
Renewable energy sources	Renewable energy sources result from the flow of energy from the Sun or Earth's interior. They include: • wind turbines (this is a variable power source, however, due to changing weather) • solar power (this can be converted into electricity using photovoltaic cells, or converted into heat energy, for example, using mirrors; like wind power, its efficiency is subject to the weather) • wave and tidal power (energy harnessed directly from the power of moving water) • hydroelectric power, or hydropower (often involving multipurpose dam schemes) • geothermal (heat from within Earth; associated with tectonic plate margins, for example Iceland).

Nuclear power   Nuclear power provides around 14 per cent of global energy but its future share fell into question after the meltdown at Japan's Fukushima Daiichi nuclear power plant in 2011.

- There are more than 400 nuclear plants around the world; 60 more are under construction and around another 500 are in the early planning stages or have been proposed.
- In 2005, four countries made up 70 per cent of global uranium production for nuclear power, including Canada and Australia.

Nuclear power stations do not emit carbon dioxide but there are significant environmental concerns about the radioactive uranium fuel and the long-term problem of disposing of nuclear waste.

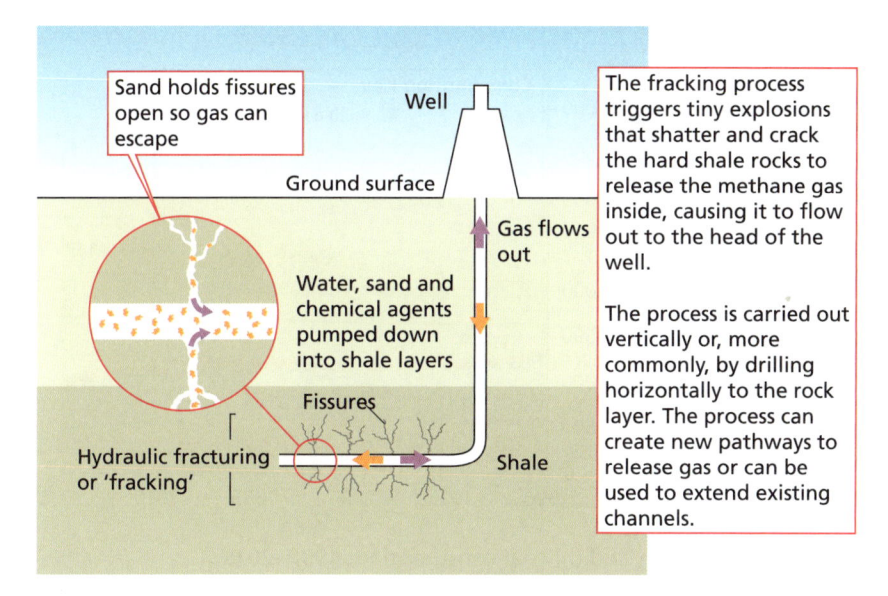

Figure 3.13 How hydraulic fracturing (or 'fracking') captures shale gas

An individual country's energy mix pattern is determined by a number of factors.

- *Physical factors* Physical geography determines whether coal or oil is found within a country's borders. Renewable energy opportunities depend on factors such as sunlight intensity, wind speeds, tidal water and deep glacial valleys (for hydropower). Table 3.6 shows how physical geography has favoured some countries with large shale gas reserves.
- *Environmental concerns* Worries about climate change have led to a growing global commitments to do more to reduce carbon emissions (see page 67). As we have seen, the Paris Agreement asks for countries that have ratified it to produce regular reports on their emissions and attempts to reduce carbon emissions (which may involve phasing out the use of coal).
- *Public perception* After the nuclear power disaster in Japan in 2011, many countries turned their backs on nuclear power temporarily or permanently. Germany and Switzerland have both announced that they will phase out nuclear power entirely.
- *Economic and political factors* The rise and fall of oil prices in recent years have had knock-on effects for the profitability of renewable energy. Meanwhile, nuclear power is back on the agenda for some European countries including the UK, owing to fears of becoming dependent on Russia for gas supplies (since the annexation of Crimea in 2014, European governments have been less keen to do business with Russia).
- *Access to technology* Renewable energy production has decreased in cost considerably in recent years, thanks to constant innovation. In some parts of the world, the costs of solar power production now compete with the costs of fossil fuel extraction and use. In the future, renewables are likely to be used far more widely as costs fall further still, as Figure 3.12 suggests.

Keyword definitions

Hydrocarbons Also called fossil fuels, these solids, liquids and gases contain a mix of hydrogen and carbon. They include coal, oil and natural gas.

Renewable energy Power sources that make use of sunlight, wind, water, vegetation and geothermal heat. Humans have burned wood for thousands of years and have harnessed water and wind power for centuries. Increasingly, however, renewables are viewed as new modern energy sources because of how they incorporate advanced technology (such as photovoltaic power cells, or turbines).

Figure 3.14 shows the global pattern of energy use for coal between 1990 and 2015. Despite its high carbon emissions, coal use soared upwards in the early years of the twenty-first century, especially in Asia. What does this suggest about the relative power of the factors driving energy choices in Asia? You can see that both production and consumption peaked in 2014 and fell in 2015. To what extent might this suggest that the world could be starting to turn its back on polluting coal in favour of renewable energy sources or cheap gas?

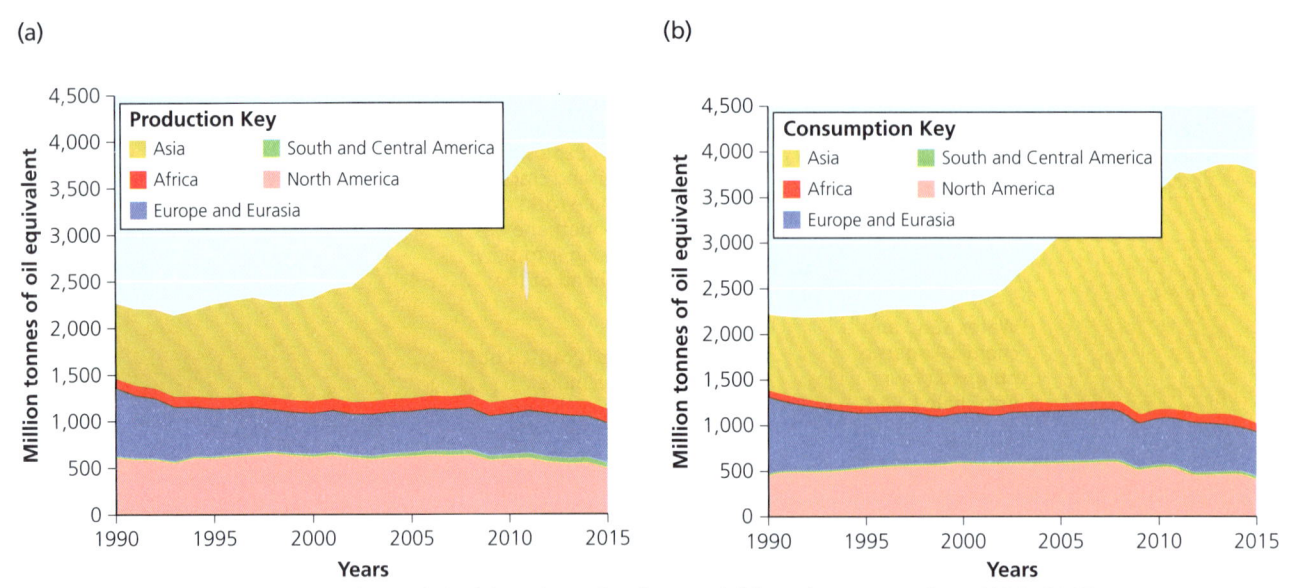

Figure 3.14 Global patterns and trends in (a) coal production and (b) coal consumption, 1990–2015

Table 3.6 Selected states with large shale gas reserves (technically recoverable resources)

Technically recoverable shale gas resources (billion cubic metres)			
Country	Reserves	Country	Reserves
Argentina	21,904	Libya	8,207
Australia	11,206	Mexico	19,272
Brazil	6,395	Poland	5,292
Canada	10,980	South Africa	13,725
China	36,082	United States	24,394

Total recoverable global shale gas resources: 187,400 billion cubic metres

■ **KNOWLEDGE CHECKLIST:**

- The concept of consumption
- Global progress towards poverty reduction
- The rise of the new global middle class (NGMC) and the fragile middle class
- The ecological footprint concept
- Global consumption of renewable and non-renewable resources
- An overview of global water consumption patterns and trends
- The concept of embedded water
- An overview of global food consumption patterns and trends
- The nutrition transition in middle-income countries
- An overview of global food consumption patterns and trends
- The changing importance of hydrocarbons, nuclear power and modern renewables

EVALUATION, SYNTHESIS AND SKILLS (ESK) SUMMARY:

- How resource consumption in different local contexts depends on spatial interactions with other places
- How population growth, poverty reduction and resource pressures are interrelated

EXAM FOCUS

DESCRIBING PATTERNS AND TRENDS

As the Exam Focus at the end of Unit 1.1 explained, your course may require you to apply what you have learned to stimulus material, such as a map or chart. Each structured question in Section A of your examination begins in this way. Below is a sample answer for an exam-style short-answer question using the command word 'describe'. Read it and the accompanying comments.

Study Figure 3.14 (page 90). Describe how the trends shown vary for (i) Asia (ii) Europe and Eurasia. [2 + 2 marks]

(i) Since the 1990s, both the production and consumption of coal in Asia has risen steeply, especially after around 2003. Production has more than doubled from roughly 1000 million tonnes to about 2500 million tonnes. The figures for consumption are very similar.

(ii) Both production and consumption have fallen in Europe and Eurasia, with most of the decline occurring in the 1990s. In both cases, the figure has declined from about 800 million to 500 million tonnes.

Examiner's comment

Both answers here are concise and well-written. The main trends are identified. Key time periods are highlighted. Data are used to support the answer (this is also called 'quantifying' your answer). The phrase 'more than doubled' is a very useful one to employ (because it emphasizes the significance of the change). Full marks would be awarded here. As practice, write your own description of changes in each of the other regions shown in the Figure.

DEVELOP THE DETAIL

Below is a sample AO2 short-answer question, and an answer to this question. The answer contains a basic point and an explanatory link back to the question, but lacks any development or exemplification. Without this, full marks are unlikely to be awarded. Write out the two basic reasons that have been given below, and then improve them by adding your own development or exemplification.

Using Figure 3.14, explain why (i) coal consumption remains very low in Africa and (ii) coal production is higher than coal consumption in Africa. [2 + 2 marks]

(i) Coal consumption is low because many people living in African countries are very poor.

As a result, they cannot afford to buy or use coal.

(ii) More coal is being produced in Africa than is being consumed.

This may be because people cannot afford to buy it, or someone else is using it instead.

How to 'max out' your marks

4 marks (2 + 2) are available here. The example answer provided would only score 2 marks (1 + 1) as it is very poor. Hopefully, the details you have added have resulted in a much better answer. Ideally, you will have named African countries in your answer (perhaps referring to a very poor country such as Mali or Somalia). You will also have identified specifically where demand for African coal may be coming from (China, perhaps). Maybe you have added some additional facts or used data: for instance, around 50 per cent of people in the Sahel are in extreme poverty (below the US$1.90 line). As a rule, use of real place names and factual evidence increases greatly your chance of being awarded extension marks in a question like this.

3.2 Impacts of changing trends in resource consumption – the Water-Food-Energy nexus

Revised

As we saw in Unit 3.1, global projections indicate that demand for freshwater, energy and food will all increase significantly over the coming decades under the twin pressures of population growth and economic development (including the growth of the new global middle class, and their nutrition transition). In Unit 2.2 we reviewed the projected effects of global climate change on water availability and crop yields: in some places water and food production will become harder irrespective of economic and demographic processes.

In recent years, the concept of a **Water-Food-Energy nexus (WFE nexus)** has emerged as a useful way to analyse the complex, mounting and interrelated pressures on our three main global resource systems.
- According to the United Nations Food and Agriculture Organization (FAO), the WFE nexus 'presents a conceptual approach to better understand and systematically analyse the interactions between the natural environment and human activities, and to work towards a more coordinated management and use of natural resources across sectors and scales'.
- According to the UK's Royal Geographical Society (with Institute of British Geographers): 'Nexus thinking draws attention to the urgent need to address the interconnected challenges of food, water and energy security, through integrated research that spans the social and environmental sciences, in ways that go beyond separate disciplinary, sectoral and policy silos.'

Figure 3.15 shows the WFE nexus: you will already recognize some of the ways in which issues affecting these three crucial natural resources are interconnected or interrelated. This chapter now explores these interactions in greater depth.

> **Keyword definitions**
>
> **Water-Food-Energy nexus (WFE nexus)** The complex and dynamic interrelationships between water, energy and food resource systems. Understanding of these interrelationships is essential if natural resources are to be used and managed more sustainably.
>
> **Safe water** Water that is safe to drink or to use for food preparation, without risk of health problems.
>
> **Food availability** Sufficient quantities of food of appropriate quality to support a population, supplied through domestic production or imports, including food aid.
>
> **Energy pathways** Flows of energy that link producer regions with consumer regions. In physical terms, pathways take the form of pipelines, electricity power lines and the routes taken by gas and oil tankers, or trains carrying coal.
>
> **Water scarcity** When the annual supply of water directly available per person falls below 1,000 cumecs.

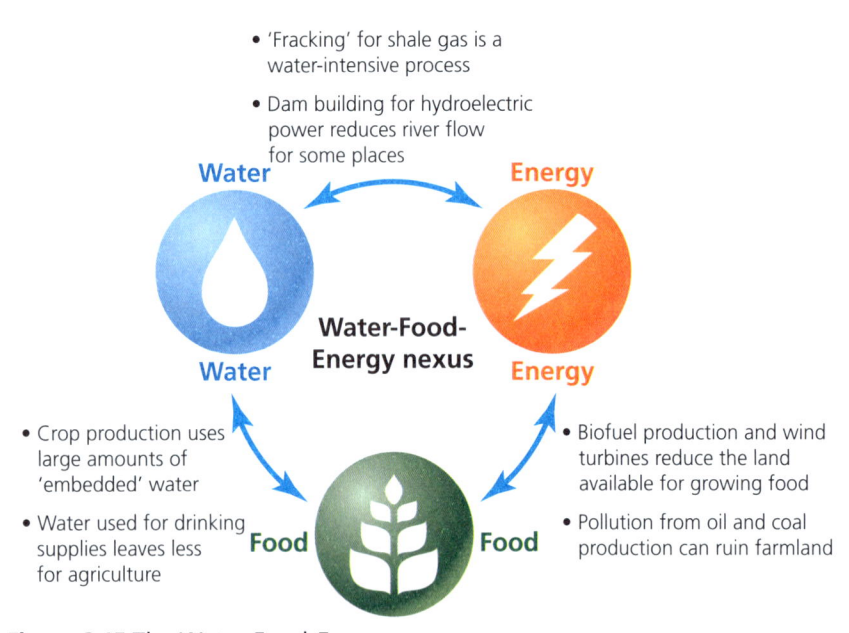

Water ↔ Energy
- 'Fracking' for shale gas is a water-intensive process
- Dam building for hydroelectric power reduces river flow for some places

Water ↔ Food
- Crop production uses large amounts of 'embedded' water
- Water used for drinking supplies leaves less for agriculture

Energy ↔ Food
- Biofuel production and wind turbines reduce the land available for growing food
- Pollution from oil and coal production can ruin farmland

Figure 3.15 The Water-Food-Energy nexus

Nexus interactions and connections

Figure 3.16 shows more of the complex interactions that require 'nexus thinking' to be applied to the national management of:
- water insecurity resulting from food and energy production
- food insecurity resulting from water and energy production
- energy insecurity resulting from food and water production.

- Agriculture accounts for 70 per cent of total global freshwater withdrawals, making it the largest user of water (Figure 3.17). Meat and dairy diets compound matters: producing 1 kg of potatoes requires 500 litres of water; 1 kg of beef requires 15,000 litres.
- 35 per cent of water used globally by the industrial sector is needed for mining, transport, processing or energy-related industries. Growth is expected in all of these sectors.
- Exploitation of domestic shale gas is seen as an important way to reduce energy security concerns by freeing sovereign states of dependency on energy imports. However, shale gas wells use vast quantities of freshwater. Estimates vary, but a typical US well uses an average of between 5 million and 8 million gallons of water. This is equivalent to a week of water use for New York city. Also, chemicals used in fracking fluids may pose health risks. In Bradford County, Pennsylvania, a major shale gas producing area of the USA, local people assert they have been poisoned and have lost access to **safe water**.
- The Guarani Aquifer underlies 1.2 million square kilometres of land shared by Brazil, Argentina, Paraguay and Uruguay. Pollution from cattle rearing enters the hydrological system and threatens the supply of safe water in multiple states.
- Large-scale water hydropower infrastructure projects can have *positive* nexus impacts by providing water storage for irrigation and urban uses. However, this might happen at the expense of downstream water users who suffer from reduced river flow.

- Some of the world's key agricultural centres are threatened by **water scarcity**. Based on their population density, vulnerable regions include: eastern parts of North America, central parts of Europe, Northern Asia, India, Southeast Asia, Japan and the Middle East.
- Some megacities that many people have migrated to are located in areas of water stress. The water requirement of many of the world's growing cities, such as Mexico City, Cairo and Los Angeles, exhaust available supplies for local agriculture.
- World production of biofuels is rising rapidly. The Earth Policy Institute estimates that land turned over to biofuels in the USA in one two-year period could have been used to produce grain for 250 million people.
- In Indonesia, huge areas of land have been taken over for palm oil production to produce ethanol (a biofuel). In addition to energy production, the TNCs who buy Indonesian palm oil, such as Unilever, use it in the manufacturing of a range of commercial food products.
- Marine pollution from oil spills can harm fish stocks resulting in food insecurity. The worst impacts are for low-income communities who are reliant on fish for subsistence needs. For example, native Alaskans who suffered major losses to their fish stocks after the Exxon Valdez oil spill in 1989.

- The food production and supply chain currently consumes about 30 per cent of total energy consumed globally according to the FAO. Energy is required to produce, transport and distribute food as well as to extract, pump, lift, collect, transport and treat water. The predicted growth in food production to meet rising consumption demands (see page 87) means that even greater amounts of energy will need to be found.
- Surface mining of coal leads to the deposition of large amounts of rock and soil waste, ruining productive land and impacting negatively on water quality once material is washed into river systems by runoff from rainfall. In excessive cases, deposition of spoil during mountain-top mining can bury streams and rivers.
- A common and serious issue associated with coal-mining all over the world is acid mine drainage, in which any mined deposit containing sulfide reacts with air and water to produce sulfuric acid. If sulfuric acid reaches rivers, a safe water supply is threatened. The damage to irrigated crops and farm animals can be very serious.
- Water safety concerns and public protests have slowed down shale gas exploration and production in Europe (fracking has been banned in Germany and France). In some people's view, this is a missed opportunity to become less reliant on **energy pathways** to the Middle East and Russia (and the geopolitical risks this dependency creates).

Figure 3.16 How national water, food and energy security issues interact with one another

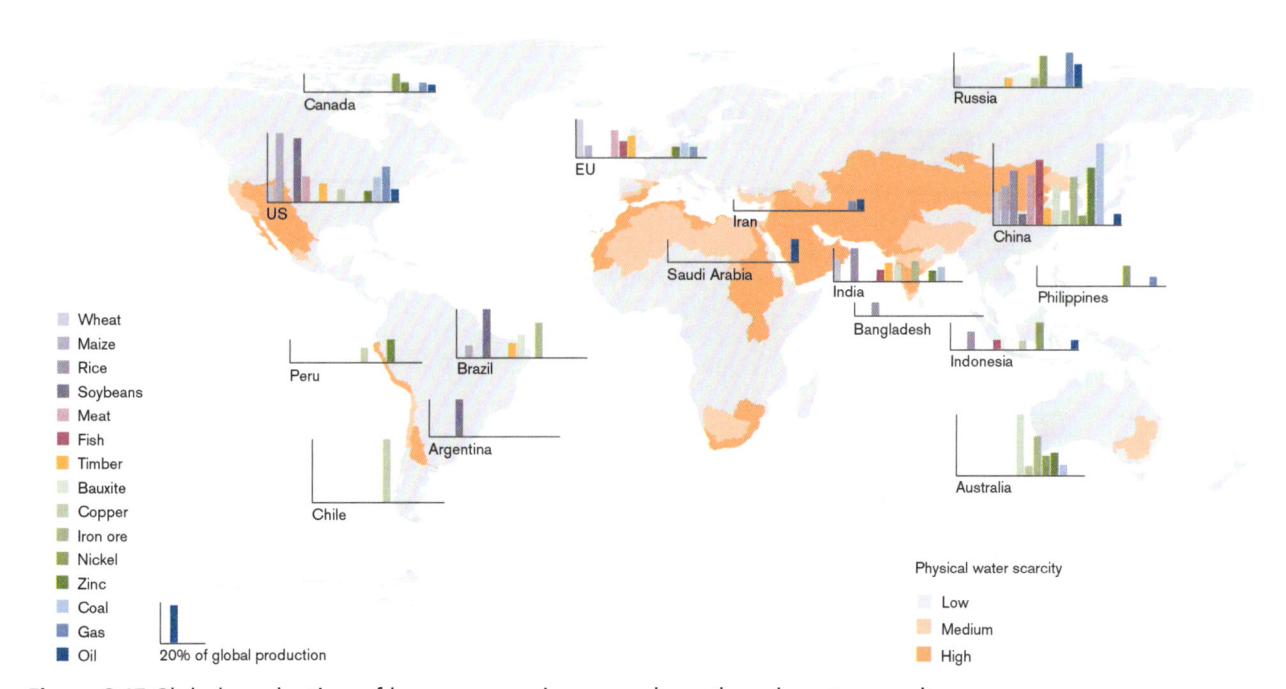

Figure 3.17 Global production of key resources in areas where there is water scarcity

The value of nexus thinking for resource management

Ideally, resource management should involve a broad range of multi-sectoral stakeholders (Figure 3.18). These may include local and national governments, development banks and agencies, international organizations, research institutes and universities, civil society organizations and TNCs. Without a high degree of cooperation between a range of powerful stakeholders, it is easy to see why some of the conflicts and tensions shown in Figure 3.16 will be almost impossible to resolve.

In its reports, the UN's Food and Agriculture Organization (FAO) uses the Red River in northern Vietnam as an example of how nexus interactions need to be factored into resource management strategies.

- A series of hydropower reservoirs in the upstream reaches of the Red River regulate water flows; they supply much of the electricity needed for Vietnam's modernization and industrialization strategies.
- However, the same river system is the sole water source for domestic uses and irrigation of almost 750,000 hectares of rice farming in the Red River delta, which is critical to social stability and food security in Vietnam.
- As water becomes scarce, and competition grows between the energy and agricultural sectors, there is still an absence of cross-sectoral consultation between different user groups.
- Nexus thinking tells us that an integrated, long-term and multi-sectoral strategy is needed desperately.

One recurring criticism of the Water-Energy-Food nexus approach, however, is that integrated management plans are not new and have not always worked in the past. Integrated water resources management (IWRM) is a well-established approach to river basin management, for instance. The FAO response to this criticism is: 'By explicitly focusing on water, there is a risk of prioritizing water-related development goals over others, thereby reinforcing traditional sectoral approaches. The nexus approach considers the different dimensions of water, energy and food equally and recognizes the interdependencies of different resource uses to develop sustainably.'

> **PPPPSS CONCEPTS**
> Think about how global flows and interactions bring environmental impacts at contrasting local and global geographic scales.

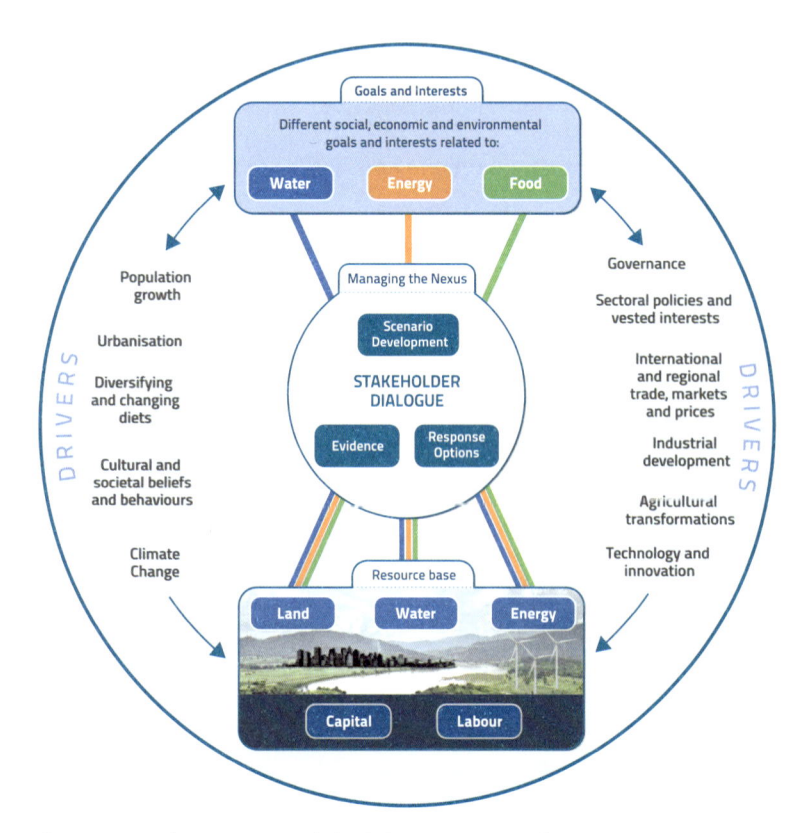

Figure 3.18 The FAO's model of the Water-Food-Energy nexus

Climate change and the Water-Food-Energy nexus

Revised

Climate change will exacerbate the challenge of meeting simultaneously the water, food and energy consumption needs of different societies. As we saw in Unit 2.2, changing rainfall and temperatures will affect biome distributions, food availability crop yields, livestock health and the spread of disease. More extreme weather events could lead to agricultural failure.

■ Future climate change and water supply

Climate change is projected to bring heightened water scarcity to many vulnerable places and societies. If extreme weather events such as drought become more frequent, desertification (Figure 3.19) and land degradation may increase in regions such as the Sahel, where rainfall has undergone a significant decrease recently.

- The Sahel region is a long strip of semi-arid drylands that borders the southern Sahara. It includes parts of Sudan, Chad, Burkina Faso, Niger and five other countries.
- Climate data for the Sahel suggest that a long-term reduction in rainfall may be taking place. Rainfall was lower in recent decades than during earlier decades of the twentieth century.
- Around 50 million poor and vulnerable people live there. There is very little money and technology available in Sahel countries to help people adapt to the challenge of shrinking water supplies.
- In turn, less water means greater food and energy insecurity: crops cannot grow, livestock may die and there is less biomass to be burned as fuel.

Other areas at high risk of heightened water scarcity include desert fringes in Australia, China and the USA (Figure 3.19). Water shortages in parts of the USA will not impact on food and energy supply in the same way as in the Sahel, however.

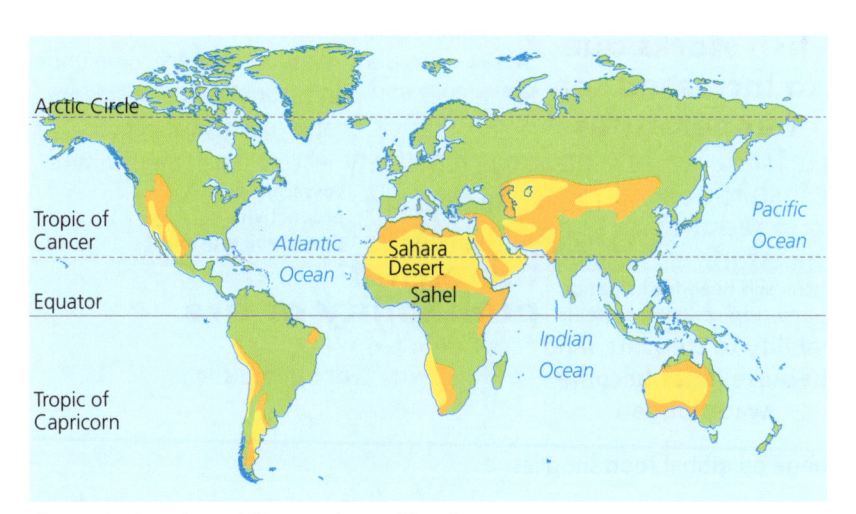

Figure 3.19 Vulnerability to desertification

■ Future climate change and food supply

Figure 3.20 shows a wide range of possible ways in which climate change might affect food supplies globally.

Rising temperatures and changes in rainfall patterns have a direct effect on crop yields, as well as an indirect effect through changes in the availability of irrigation water or an increase in the frequency of wildfires. Livestock farming in the Sahel, the fringes of the Australian desert and other semi-arid regions will be affected in terms of greater water requirements for livestock and livestock heat stress. The FAO is concerned this could trigger a shift from transitory food insecurity to chronic insecurity for some societies.

- *Transitory food insecurity* is short-term and temporary. Many people living in semi-arid regions naturally experience short-term fluctuations in food availability and food access. This is due to the highly seasonal nature of agriculture in parts of the world where no rain falls for long parts of the year. The Kishapu District in Tanzania gets very little rain, for instance, but has fertile land. During the wet season, farmers grow crops that take a short time to mature, such as sorghum, finger millet and sunflowers. With careful planning, potential food insecurity can be can be overcome: surplus harvests are stored for use in the dry season.
- *Chronic food insecurity* is long-term or persistent. People are unable to meet their minimum food requirements over a sustained period of time. Climate change threatens places like Kishapu District with this more chronic form of food insecurity. Longer periods of drought mean that natural resources in this fragile environment will lose resilience: soil erosion and the permanent loss of vegetation may bring desertification. Any reduction in the growing season reduces farmers' ability to generate a food surplus at harvest time. Without this, communities will not be able to survive the dry season.

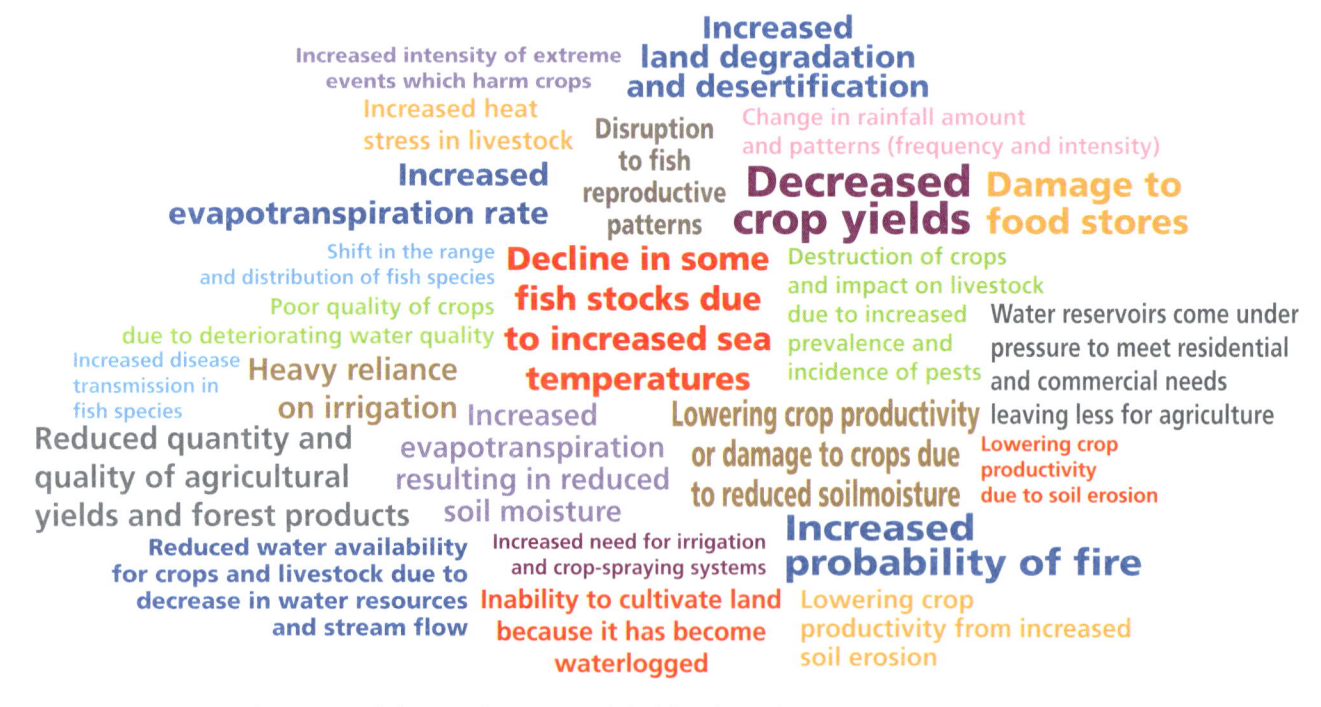

Figure 3.20 Potential impacts of climate change on global food supplies
Source: WWF

Future climate change and energy supply

There will be many direct effects of climate change on renewable energy production.
- Shrinking ice stores and changing river regimes may threaten hydroelectric power supplies. Power generation companies in China currently make use of massive seasonal runoff from the Himalayas and the Tibet plateau. However, climate change threatens to permanently reduce the size of glacial ice stores in the region. In the long term, it could lead to dangerous water and power shortages – because there will be very little ice left to melt.

- Changes in wind patterns, cloud cover and rainfall may impact on renewable energy production using wind or solar power (however, there may be both positive and negative effects). Biofuels may become harder to grow in some places unless additional irrigation water is provided – with knock-on effects for water security.

Fossil fuel energy production is a nexus element that is less clearly affected by climate change: our ability to mine coal or use existing gas and oil wells is not impacted on in any obvious negative way by temperature rises. In fact, global oil reserves may increase in size on account of improved access to Arctic Ocean oil and gas fields as a result of sea ice melting (see page 57).

- Figure 3.21 shows oil and gas fields under the Arctic Ocean: 30 billion barrels of recoverable oil and many trillion cubic metres of gas are believed to lie beneath the Beaufort Sea and Arctic Ocean. TNCs like Shell, Chevron and Statoil began exploratory undersea drilling there in 2012.
- Although low world oil prices since 2014 have made offshore oil relatively expensive and unprofitable, pressure to exploit Arctic oil may grow in the future.

Climate change does bring *political* uncertainty over the continued use of fossil fuel energy supplies, however.

- Climate change mitigation – including potentially binding global agreements to reduce carbon emission (see page 68) – may compel governments to invest more money in renewable energy while curbing fossil fuel use (which could be done by introducing a carbon tax, or passing laws to prevent exploitation of Arctic oil and gas).
- One argument that is sometimes used to support shale fracking is that gas is a cleaner energy source than coal in terms of carbon emissions. Governments may therefore become more willing to permit fracking, despite the alleged threat to water security that this brings.

One final element of uncertainty that climate change introduces into energy supplies is the way in which hotter temperatures could increase the demand for air conditioning or desalinized water – both of which put higher pressure on energy supplies.

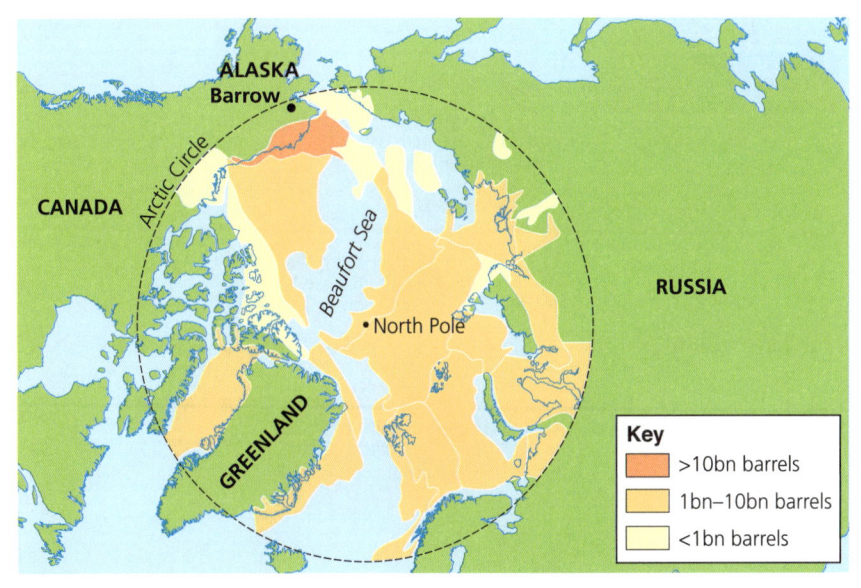

Figure 3.21 Oil and gas fields under the Arctic Ocean

CASE STUDY

WFE RESOURCE INSECURITY IN INDIA

Few regions are more environmentally insecure than south Asia. The region faces rising sea levels and regularly experiences coastal flooding already — which is of particular concern in a region with heavily populated coastal areas. India – which houses nearly a fifth of the world's population, yet contains less than 5 per cent of its annual renewable water resources – is viewed as having high vulnerability to, and limited resilience against, more severe IPCC climate change projections (Table 3.7). Figure 3.22 shows some of the processes of change which threaten the country's food security.

Table 3.7 Future nexus possibilities for India

Population trends	■ India is home to over 1.2 billion people and is predicted to be the world's most populated country by 2022. It has not competed demographic transition yet and has a rate of natural increase of 11/1,000/year. This means there could be 1.7 billion people living in India by 2050, by which time the country may be experiencing water and food security challenges due to a changing climate.
	■ Population density in India is 380 people per square kilometre, which is considerably higher than the average population density of the world as a whole.
Water possibilities	■ 250 million people currently lack access to clean water; trying to improve their situation will increase the pressure on parts of the country where water is already in scarce supply.
	■ India's total demand for water is expected to exceed all current sources of supply and the country is set to become water scarce by the year 2025 (measured as per capita water availability of less than 1,000 cubic metres annually).
	■ The situation will worsen even further if ice stores in the Western Himalayas – which provide water supplies to much of India and south Asia – become permanently reduced in size due to climate change.
Food possibilities	■ Because of rising temperatures, India's agricultural productivity could fall by 35 to 40 per cent by 2080. Of particular concern is the productivity of the Punjab region in northeast India, which currently produces large amounts of the country's wheat and rice but has suffered frequent droughts in recent years.
	■ Food insecurity problems will manifest most acutely in India's least developed and poorest regions and may exacerbate conflicts between different social groups.
Energy possibilities	■ 400 million Indians still do not have electricity. One way of solving this problem is to adopt greater use of hydropower – India has recently increased its construction of hydropower projects on the Indus Basin's western rivers. However, these waters are allocated to Pakistan by the Indus Waters Treaty. Rising tensions between the two states may escalate if water flows are reduced as a result of climate change.

Figure 3.22 Food security under threat in India

Factors surrounding India's threatened food security:
- Urbanisation reduces availability of agricultural land
- Growth of middle class increases resource pressure
- Population predicted to increase from 1.2 billion to 1.6 billion
- Water used by industry leaves less for agriculture
- Irrigation threatened by hydropower dams
- Increased soil erosion due to intense rainfall
- Fall in groundwater levels
- Increased frequency of floods and droughts
- Unreliable weather patterns and monsoon

3.2 Impacts of changing trends in resource consumption – the Water-Food-Energy nexus

CASE STUDY

WFE RESOURCE SECURITY IN CANADA

All states face WFE security challenges on account of climate change, some of which may be linked to the physical threats of sea-level rise or increased extreme weather events. Compared with many other countries, however, Canada is viewed widely as having relatively low vulnerability. Models suggest that Canada's major cities are among the world's most resilient when it comes to future climate change, though Vancouver's low-lying coastal location makes it relatively vulnerable to sea-level rise.

Figure 3.23 demonstrates that large habitable areas of Canada are under-populated currently; in the future, the carrying capacity of large areas of Canada will be able to support greater not fewer numbers of people if GMST continues to rise. Table 3.8 shows future nexus possibilities for Canada.

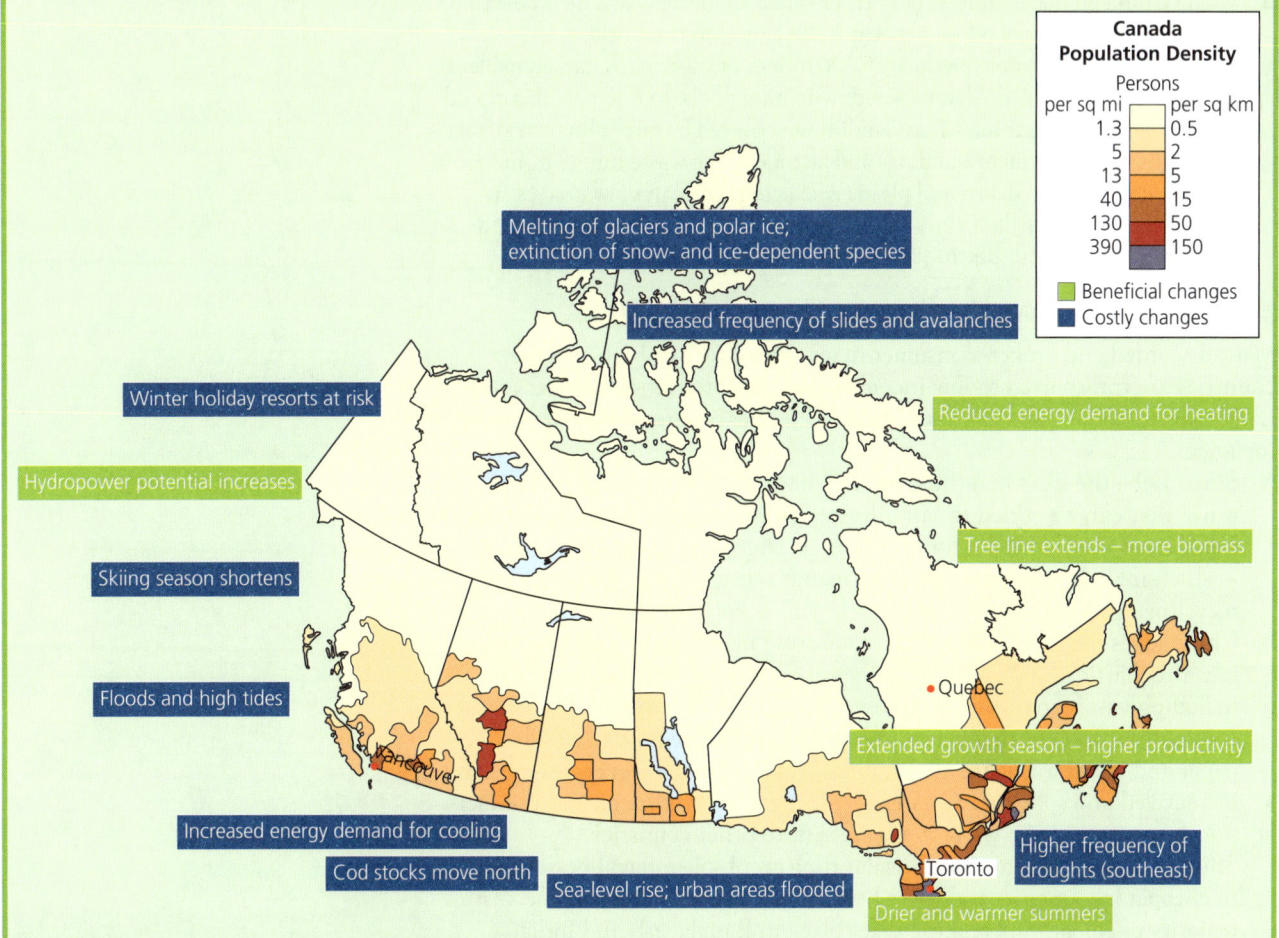

Canada Population Density

Persons per sq mi	per sq km
1.3	0.5
5	2
13	5
40	15
130	50
390	150

- Beneficial changes
- Costly changes

Costly changes:
- Melting of glaciers and polar ice; extinction of snow- and ice-dependent species
- Increased frequency of slides and avalanches
- Winter holiday resorts at risk
- Skiing season shortens
- Floods and high tides
- Increased energy demand for cooling
- Cod stocks move north
- Sea-level rise; urban areas flooded
- Higher frequency of droughts (southeast)
- Drier and warmer summers

Beneficial changes:
- Reduced energy demand for heating
- Hydropower potential increases
- Tree line extends – more biomass
- Extended growth season – higher productivity

Source: http://media.web.britannica.com/eb-media/10/70010-004-AFC7AD58.gif

Figure 3.23 Population distribution in Canada and some projected effects of climate change

Table 3.8 Future nexus possibilities for Canada

Population trends and patterns	■ Just 35 million people live in the world's second-largest country, resulting in a population density of less than 4 people per square kilometre. In stage 4 of the demographic transition model, Canada's population will not increase greatly either. ■ Vast areas of Canada are unpopulated and some regions may become more rather than less habitable. While some coastal land may be lost, Canada does not face the same serious threat of losing large areas of habitable land that many other states face.
Water possibilities	■ Canadian people will not suffer from water scarcity: with such a small population and vast land area, any locally occurring water shortages can always be tackled using transfer schemes from other parts of the country.
Food possibilities	■ In parts of Canada, the annual mean temperature has increased by 1.5°C since 1900. Further rises in temperature will increase the length of the thermal growing season and opportunities for agriculture in some regions of Canada. For instance, a 25–30 per cent increase in potato and wheat yields is estimated at some high latitudes. Productivity improvements in northern regions could reach 40–50 per cent by the 2080s. ■ A likely consequence of warmer waters is a change in the composition and abundance of fish species in different sea areas. In some areas the commercial value may be reduced, while in other areas it might increase.
Energy possibilities	■ Opportunities for hydropower may increase with higher rates of ice melting due to climate change. At the same time, demand for electricity may decrease due to warmer temperatures. The net result may be greater rather than lessened energy security. ■ Melting sea ice may open up opportunities for oil and gas extraction in Canadian Arctic waters.

Waste disposal and recycling

The ecological footprint calculation (page 83) gives consideration to the amount of land and resources required to dispose of our waste. The more levels of consumption rise, the more waste there is to tackle. In emerging economies, waste disposal has become a major challenge, especially for megacity planners (see page 19).

- Thanks to rising incomes and higher standards of living, China generates waste at a rate that has risen twice as fast as population growth over the past several decades.
- While China produces far less rubbish per capita than the USA, its production of 400 million tonnes of refuse per year is the largest in the world.
- Beijing's 22 million people produce 25,000 tonnes of waste each day, equivalent to the weight of 10 filled Olympic-sized swimming pools. In one neighbourhood of Beijing called Tongzhou, a huge landfill site covered by tarpaulins rises like a mountain amid apartment buildings and factories. The waste ranges from food waste to construction debris and plastic packaging. You can view images at: https://www.theguardian.com/environment/gallery/2010/mar/26/beijing-rubbish-wang-jiuliang-photography

International waste flow patterns and issues

Many discarded and broken consumer items originating in high-income countries are transported to low-income countries for disposal and recycling. Large international flows of waste connect European states with Ghana, for instance.

- Since 1999, the EU's Landfill Directive has reduced the proportion of waste that can be buried in landfill sites. In order to meet targets, greater amounts of waste are sent for recycling. A large proportion of this waste – which includes paper, plastics and metals – is exported overseas for recycling.
- China leads in global waste trade, importing more than 3 million tonnes of waste plastic and 15 million tonnes of paper and cardboard a year, including vast amounts coming from the EU. In this context, however, it is important to note that waste streams are profitable for China; plastic, paper and metal waste are all resources which, if processed correctly, can be recycled and sold as raw materials for the manufacture of new products.
- Sometimes, very large waste items are sent to other countries. Shipbreaking is the process of dismantling an obsolete naval vessel. Owing to cheaper labour costs and fewer health and safety regulations, the vast majority of global shipbreaking takes place in Bangladesh and India.
- Toxic waste was moved illegally from Europe to Ivory Coast in 2006. Tens of thousands of Ivorians suffered ill health after toxic waste alleged to produce hydrogen sulfide was dumped by a ship in the employ of Trafigura, a European TNC. A long legal battle to win compensation for the victims resulted eventually in a US$46 million cash settlement shared by all those affected.

CASE STUDY

INFORMAL E-WASTE RECYCLING IN GHANA

E-waste – electrical and electronic waste – is one of today's fastest-growing transnational waste streams. By managing it well, society can recover valuable raw materials and reusable parts, with significant associated carbon dioxide emissions savings (compared with using new natural resources). E-waste includes every bit of abandoned electronic and electrical material belonging to computers, CD players, smartphones and printers. Normally, the parts are dismantled and the waste burned to extract expensive metals, including gold, silver, chromium, zinc, lead, tin and copper. Large numbers of people in India and China, including children working in family-run workshops, take part in informal e-waste recovery. Some of the worst and most widely reported problems have arisen in Ghana, however.

Electronic waste imports into Ghana exceed 200,000 tonnes annually, of which 70 per cent are second-hand goods, including donations of old computers sent to schools by charities. However, around 15 per cent of second-hand imports will turn out to be broken beyond repair. Additionally, many non-functioning and non-reusable electronic devices are sent there deliberately from the EU (despite European laws that are supposed to prevent this).

There are no distinct policies in Ghana for the regulation and management of e-waste. As a result, a large informal industry has developed around the recovery of e-waste, but there are no clear regulations to prohibit unsafe work practices.

- The settlement of Agbogbloshie or Old Fadama consists of about 6,000 families or 30,000 people, situated on the left bank of the Odaw River in Accra. The establishment and growth of Agbogbloshie is driven by poverty and the migration of desperate people from the north of Ghana, some of whom are escaping tribal conflict. Over the years it has grown into a dumping ground for old electrical and electronic products. Hundreds of tonnes of e-waste end up there every month as a final resting place, where they are broken apart to salvage copper and other metallic components that can be sold.

- In Agbogbloshie, entire families work in appalling conditions. The method of extracting and recovering valuable materials from old computer circuit boards is highly hazardous (Figure 3.24). The burning process releases toxic substances into the atmosphere, soils and water, with dire health consequences. Known health problems include acute damage to the lungs from inhalation of fumes of heavy metals such as lead and cadmium.

Figure 3.24 Working with e-waste in Accra, Ghana

- Toxic wastes, heavy metals and battery acids released into the soil and the surface water have destroyed wildlife in the Odaw River, which used to be an important fishing ground for the neighbouring communities.

■ KNOWLEDGE CHECKLIST:

- Interactions occurring within the Water-Food-Energy (WFE) nexus
- How the WFE nexus can affect water security and safe water access
- How the WFE nexus can affect food security and availability
- How the WFE nexus can affect energy security and energy pathways
- The value of nexus thinking for resource security
- Future climate change and water supply
- Future climate change and food supply
- Future climate change and energy supply
- Contrasting examples of how climate change could affect national resource security (India and Canada)
- Waste disposal and recycling issues
- International waste flow patterns and issues
- Informal e-waste recycling in Ghana

EVALUATION, SYNTHESIS AND SKILLS (ESK) SUMMARY:

- How different WFE nexus elements interact spatially with one another
- How perspectives may vary on which resource security issues are most important

EXAM FOCUS

SYNTHESIS AND EVALUATION ESSAY-WRITING SKILLS

Below is a sample answer to an exam-style essay question (see also pages 74–75). Section C of your examination offers a choice of two essay questions for you to choose from. Having made your choice, you need to plan carefully an essay response which **synthesizes** and **evaluates** information (under assessment criterion AO3). You can draw on ideas from any of the chapters in this book. This means that as you progress through the course, a greater volume of ideas, concepts, case studies and theories become available for you to synthesize. This sample answer draws on ideas from all three units of this course.

The question has a maximum score of 10 marks. Read it and the comments around it. The levels-based mark scheme is shown on page vi.

To what extent can the Water-Food-Energy nexus concept help societies to improve how they manage their natural resources? [10 marks]

The Water-Food-Energy nexus concept basically involves thinking about the way that water, energy and food resources are interrelated. For instance when a society makes high demands on agriculture there are also high demands on water supplies and energy. This is because water is needed to grow the food and energy is needed to transport it to market. (1)

Thinking about natural resources in this way can be very useful when a society is planning for sustainable development. A good example of this is the Red River in Northern Vietnam where the government is building a series of reservoirs. The water will be used to power turbines and generate hydroelectricity. However, doing so also creates huge problems for food supplies because many of the areas of rice farming lower down in the Red River delta do not receive as much water now that it is trapped in the dam. What this shows is that nexus thinking was needed at the planning stage when the dams were being built, and perhaps the dam designers should have thought more carefully about the needs of the rice farmers. (2)

Many people believe that climate change is going to place more stress on water, food and energy supplies in many parts of the world. The IPCC has predicted that drought could increase in some parts of the world such as Arizona in the southern USA. This will create huge problems for water supplies to cities like Phoenix. It will also create problems for agriculture and possibly hydroelectric power, if dam levels in the river Colorado fall too low. Therefore it is vital that planners use the nexus concept and realise that water supply shortages will have knock-on effects for food and energy sectors as well. (3)

On the other hand, it can be argued that the nexus concept does not always help societies to manage their natural resources more sustainably. In any situation the needs of either water, food or energy producers may take priority over everyone else because of their financial or political power. This happens in some parts of the world where powerful TNCs divert water supplies for their food crops and do not really care about what the implications are for local people's water supplies. This happened in Kerala in India. (4)

It is also difficult to get different countries to work together towards nexus management goals. Not only do the needs of water, food and energy producers in one country like the USA all have to be satisfied, so too do all the different user groups in Mexico, because they are all sharing the same water supply, i.e. the River Colorado. In reality it could be very hard to arrive at a working agreement that makes everybody happy. (5)

1. This first paragraph provides a reasonable definition of the Water-Food-Energy nexus in excess, although it is a pity that natural resources are not also defined, as this is the other key concept which features in the essay title.

2. This is a useful and broadly accurate case study that shows why Water-Food-Energy nexus thinking is needed. It would be interesting to know what, if anything, could have actually been done at the planning stage to accommodate the rice farmers while also meeting the needs of the energy sector. Ideally, we need to know how helpful the Water-Food-Energy nexus concept has been in practice (rather than simply as a theory or model).

3. This paragraph also uses evidence well to exemplify the issues for the USA while linking nexus thinking with the topical issue of climate change.

4. In this paragraph, the answer begins to address the other side of the argument. This suggests the essay has been well planned. Use of the phrase 'on the other hand' helps the reader know that we are now addressing the counter-argument. This paragraph is in danger of becoming over-generalized but luckily a brief example is provided latterly which helps to strengthen the argument. It is also good to see the key concept of power being used and there is understanding demonstrated of the way that nexus thinking may not prevail if powerful forces pursue their own selfish ends.

5. Another good counter-argument is raised here, which draws on the concept of geographic scale (even if the word 'scale' is not actually used). Here, the writer of the essay grasps well the complexity of dealing with the needs of user groups *within* a country and also *between* countries.

Examiner's comment

It is a pity that there is no conclusion or final judgement included as part of this essay. It would be unlikely to score full marks as a result, because we do not know the extent to which the writer ultimately agrees or disagrees with the title.

Overall, the use of evidence and the balance between arguments both in favour of and against the statement mean that this answer would certainly reach the penultimate mark band of seven or eight marks. Can you write a conclusion which follows on from this essay that might help it to advance to the top band of 9 or 10 marks? Aim to write no more than 100 words. You should review both sides of the argument briefly and make a final comment on the helpfulness or value of the nexus concept.

Content mapping

You have now almost completed the course and there is a large amount of material that can be drawn on to help answer this essay. This answer has made use of case studies, concepts, theories and issues drawn from all of the first eight chapters of the book. Using the book's contents page, try to map the content.

3.3 Resource stewardship possibilities

Revised

Can Earth cope with more people? When the 7 billion milestone was passed in 2012, media reports were mixed in their views. Some writers were concerned and pessimistic. Others were relaxed and optimistic. A few stressed the urgent need for natural resource to be managed using **stewardship** principles at both local and global scales (Figure 3.25). Divergent views had been aired similarly in 1999 when the 6 billion milestone was reached, and in 1987 when Earth's population first reached 5 billion (Figure 3.26). In fact, people have held diverging beliefs about the costs and benefits of population growth for hundreds of years. This section begins by reviewing a debate that first began in 1798.

Keyword definition

Stewardship An approach to resource management which views humans as 'caretakers' of the natural world.

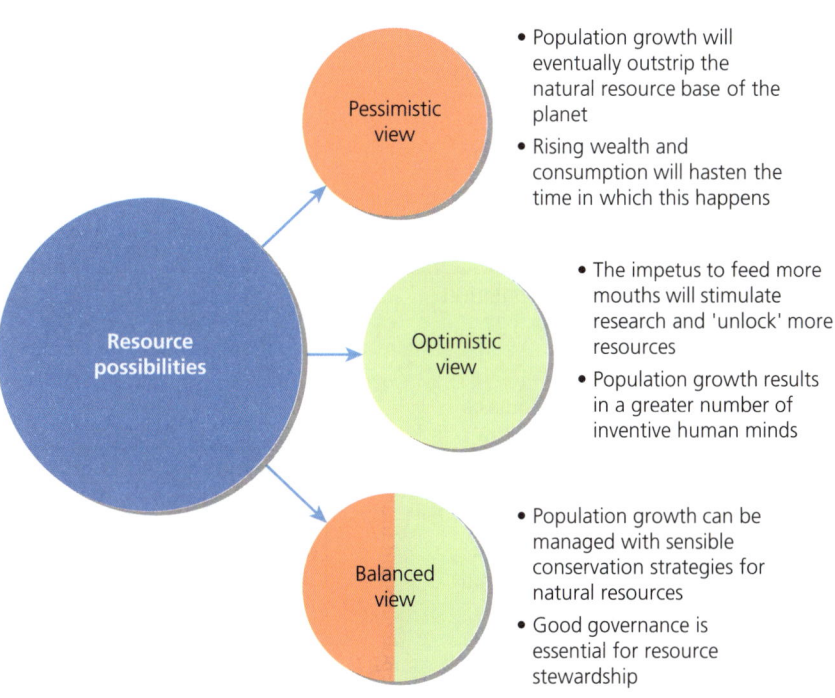

Figure 3.25 Three diverging views on the possibility of sustainable natural resource management

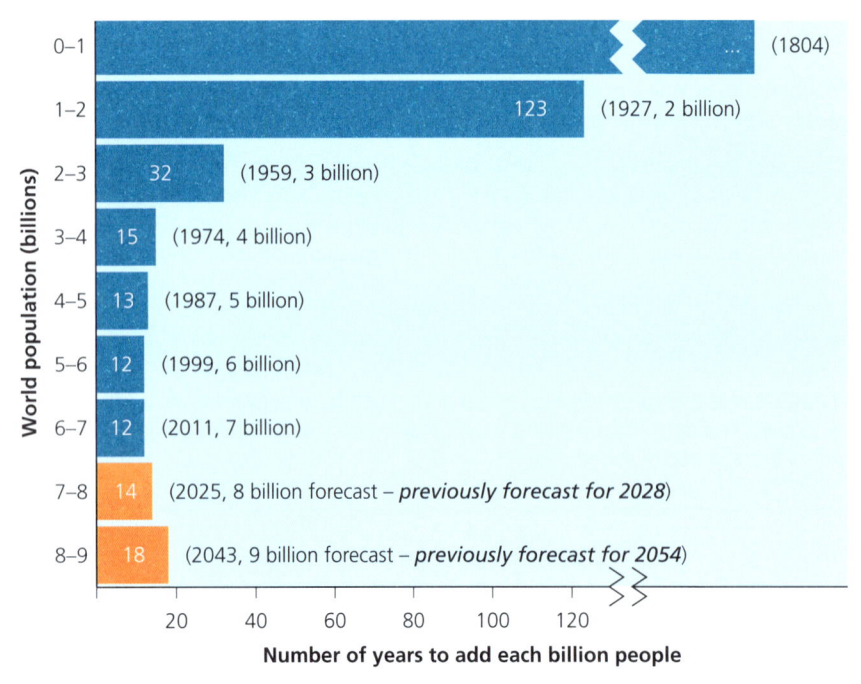

Figure 3.26 Population size milestones and the time it took to reach them

Pessimistic views on population growth

Revised

Modern study of possible relationships between population and resources can be traced back to Thomas Malthus (1766–1834). The English author of the essay *Principle of Population* (1798) was convinced that population growth was a dangerous process. He argued there was a fixed environmental 'ceiling' to population growth and that a growing number of people would outrun their food supply ultimately in the absence of any constraints (or 'checks') on their reproduction. Malthus maintained that:

1. A human population can potentially grow at a geometric rate, doubling every 25 years or so. Two parents have four children, eight grandchildren, and so on (i.e. the population growth trend is 2, 4, 8, 16, 32...).

2. Improvements in food production from the land could, at best, increase at an arithmetic rate, in the view of Malthus (i.e. 1, 2, 3, 4, 5...).

3. Very quickly, according to this argument, the ratio between population and food supply will become too large for a disaster of some kind to be avoided. No matter how few people there are to begin with, the exponential population growth curve will eventually intersect the arithmetic food line. Figure 3.27 shows population growth finally overshooting the food supply.

4. Once 'overshoot' has occurred, the growing lack of food *per capita* becomes a 'positive' check on further population growth: some combination of famine, disease and war is inevitable, according to Malthus. This will bring population numbers back to sustainable levels.

5. Alternatively, 'negative' (preventive) checks might be introduced before the ceiling to growth is reached. Positive checks can be avoided if, for example, a society adopts a later age of marriage, which would reduce its fertility rate.

His arguments were grounded in evidence. Firstly, Malthus had observed personally a close correlation between fluctuations in crop prices and marriage rates in rural England (when prices were low, farming incomes fell; people delayed getting married and having children as a result). Secondly, there were clear limits to agricultural productivity gains in the early 1800s (when agricultural science was still in its infancy). Malthus's pessimistic argument was also informed in part by the economically and politically unstable times he lived through. In neighbouring France, food shortages played an important role in the violent revolution that had just taken place.

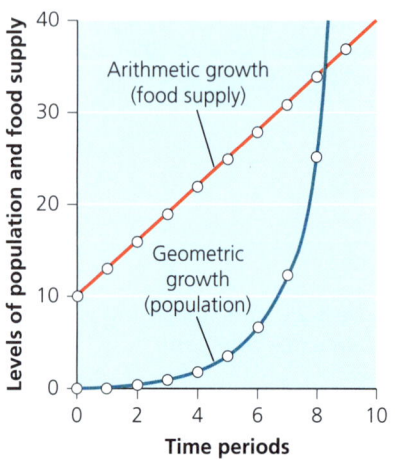

Figure 3.27 The classic 'Malthusian equation'

■ Neo-Malthusian views

More than 200 years have passed since Malthus wrote his famous essay and its influence is still felt strongly. Ideas and concepts introduced by Malthus are used widely by academics and journalists alike in contemporary accounts of famines and population polices. For instance, China's one-child rule (page 31) is often described as an example of a 'preventive check' on population growth.

Two especially influential twentieth-century books adopted and updated the framework Malthus introduced. *Limits to Growth* (1972) and *Beyond the Limits* (1991) were written by a group of academics calling themselves the Club of Rome.

- They argued that the world is fast approaching a point where a positive check to further growth is inevitable.
- Applying a conceptual approach very similar to the ecological footprint (see page 83), the Club of Rome used computer modelling of the world system to conclude that humans will soon overshoot the carrying capacity of the environment at a global scale. There is certainly plenty of evidence, included in Units 2 and 3 of this book, that might help support such an argument.
- Figure 3.28 shows the Limits to Growth projections. As you can see, it incorporates the geometric or exponential growth rate of population suggested by Malthus. For this reason, the Club of Rome's work has been labelled as 'Neo-Malthusian'. The Limits to Growth model also includes new features. You can observe the pollution increases that are anticipated when an increasingly resource-deprived population is forced, finally, to use even the most impure and polluting minerals, ores and fuels.

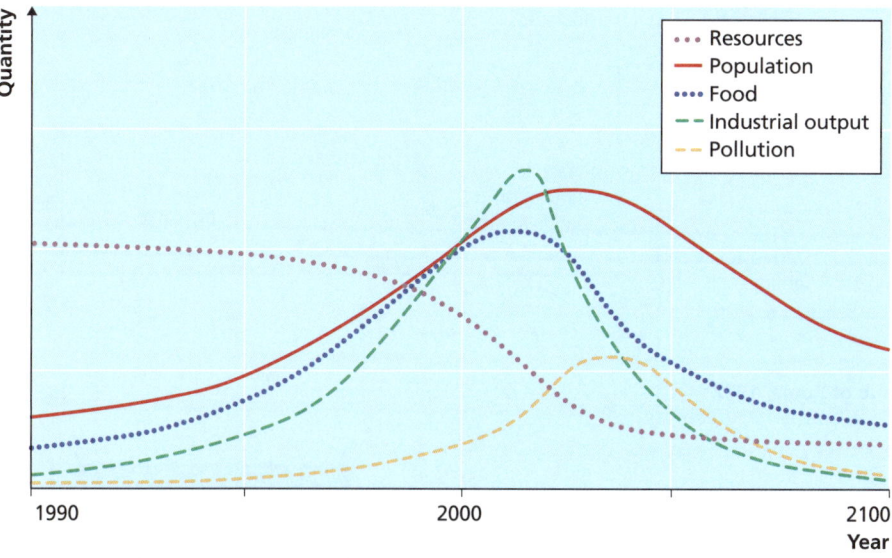

Figure 3.28 The Limits to Growth model (a pessimistic projection that shows world population falling once resource availability collapses in the 2020s–30s)

Optimistic views on population growth

Revised

Both Malthus and the Neo-Malthusians can be criticized for narrow thinking. From a radical or Marxist economic perspective, food shortages are less likely to arise from ecological limits than they are from unfairness in the way natural resources are owned, managed and consumed. Some of the data shown in Figure 3.4 (page 78) support this view. From a philosophical perspective, Julian Simon's arguments (page 33) can be used to reject the portrayal of people as a 'problem' by the Neo-Malthusians.

Writing in the 1960s, Ester Boserup produced a thesis which many people view as the ultimate anti-Malthusian theory. While Boserup did not set out to attack Malthus directly, implicit in her work are many points that can be used to build a case against Malthus and the Club of Rome.

- Boserup's optimistic theory claims that population growth stimulates resource development, and not the other way around. Instead, it is the drive to feed more mouths that stimulates the scientific community into working to raise the carrying capacity of their environment.

- Boserup's own research work involved making anthropological studies of subsistence farming improvement in very poor countries during the 1960s. Her convincing field evidence suggests that population growth encourages crop rotation and changes in land ownership, which increase greatly the efficiency of land use. Crucially, such reforms occur *only when the threat of population pressure is evident and the carrying capacity of the land is about to be exceeded*. Her argument is likened often to the old saying that 'necessity is the mother of invention'.

Does the evidence support the Boserup thesis?

Over time, Boserup envisaged a series of sudden leaps forward in the scale of food production, each triggered by the necessity of accommodating population growth. In contrast, Malthus and the Club of Rome foresaw food and resource shortages arising because of society's inability to innovate at a rate which can keep up with population growth (Figure 3.29).

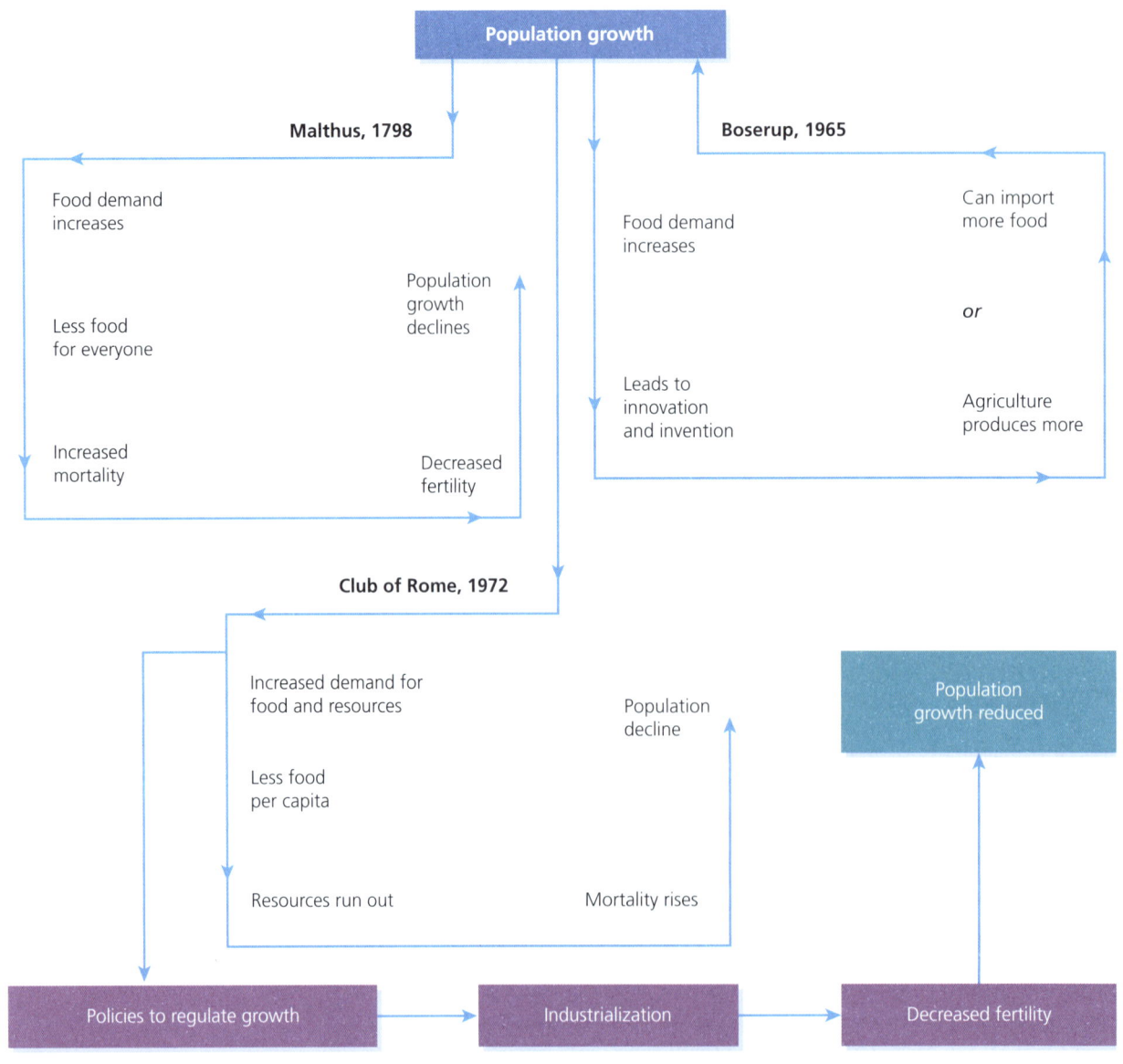

Source: Mike Witherick

Figure 3.29 Diverging theories about the relationship between food/resource availability and population growth

History does show that Malthus was wrong. Food supply growth actually outstripped population growth globally during the 1800s and 1900s. In many of the most densely populated nations of the world, successful use of technology has raised the carrying capacity of the land via three routes:

1 *Intensification* of agriculture involves raising the yields of existing land, perhaps through crop rotation, irrigation, drainage and land reform.

2 *Extensification* of agriculture involves developing remote areas for farming when transport infrastructure improves (as happened in North America during the early 1900s).

3 *Technological developments* include (1) pesticides and fertilizers, (2) the development of high-yielding cereals (IR8, the world's first high-yielding rice, is a famous example) or livestock, and (3) mechanized techniques (which have increased milk and egg production dramatically).

However, some people dispute the apparently relaxed attitudes of population growth optimists in the twenty-first century. As we have seen, the climate change and WFE nexus challenges that lie ahead are enormous (and some would say insurmountable). The new context of rising consumption by the NGMC must be factored in too ('classic' population theories and models pay far more attention to changes in a population's size than they do its wealth).

Balanced views on population growth

Is 9 billion people too many? Or 11 billion, should it come to that? There is no simple answer to that question because so many different interrelated issues and dynamics are involved (Figure 3.30). Along with the WFE nexus and climate challenges, these also include changes in population size and structure, technological innovation, economic development processes and the changing aspirations of men and women as they escape poverty.

On the one hand, history offers hope by demonstrating repeatedly how human ingenuity has helped us thrive in difficult places and at tough times: 9 (or 11) billion people will provide a very large number of minds. Surely there will be enough brainpower to conjure up whatever technological fixes the future requires?

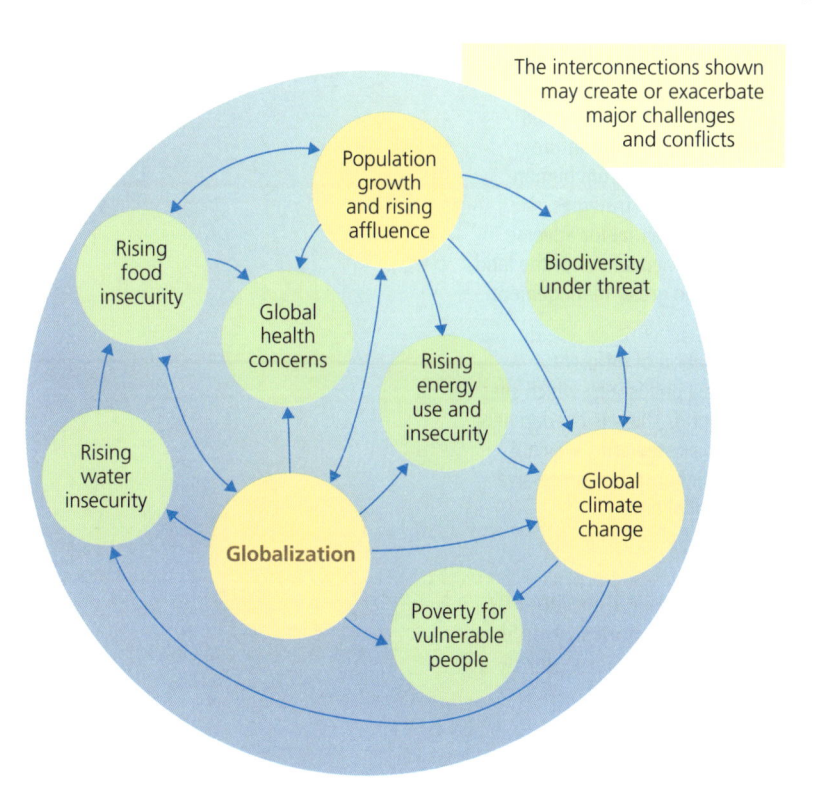

Figure 3.30 Interrelated issues that may help determine future population growth possibilities

Improvements in the efficiency and cost of solar power, for instance, could hold the key to wide-scale desalinization of seawater, which would unlock in turn the agricultural food-producing potential of vast areas of desert.

Yet on the other hand, signs have arrived already that we are on the cusp of an unsustainable future: the early warning signals of sustained rises in GMST, melting ice and global sea level (see Unit 2.2) cannot be ignored.

A pragmatic, balanced response to the uncertainties ahead could involve doing more to conserve resources, rather than continuing with 'business as usual' while relying on future technological fixes such as carbon capture and improved solar power. Improved governance of the natural resources we already use could go a long way towards improving global prospects for a sustainable future.

- More could be done to address the *equity* of natural resource availability and consumption. Global food production continues to keep pace with population growth, for instance: it is just distributed unevenly. In a single day, one North American consumes 5 kilograms of food compared with the 0.5 kilograms that are available to an average sub-Saharan African. The problem, it can be argued, is not that there are too many people in the world but that human society does not think and act yet as a unified global community: food and resources are still not being shared equally and fairly.
- More could be done to reshape the relationship between humanity and the natural world. Powerful governments, organizations and businesses need to act as 'stewards' by adopting conservation and preservation principles in order to safeguard natural resources for the future (see Table 3.9). There is no reason why this cannot be done while also promoting continued economic development. As pages 100–101 explain, economic systems are currently extremely wasteful and much can be done to streamline their efficiency.

Table 3.9 The stewardship approach to managing natural resources combines conservation and preservation strategies

Conservation	Conservation involves the efficient and non-wasteful use of natural resources. It is an important foundation concept for sustainable management (or sustainable development). This is because it acknowledges that humanity has an obligation to consume resources in an efficient way that does not leave its descendants deprived of things they will need to survive.
Preservation	While conservationists view the environment as a source of natural resources that can be made available for (sustainable) commercial exploitation, a preservation management approach views nature and especially wilderness as something best left apart from human commerce. National Parks and so-called 're-wilding' strategies have their roots here. However, there is strong potential for adverse economic effects on people. This is because commercial use of the land could be banned altogether under a 'keep off the grass' management approach.
Stewardship	Environmental stewardship involves a combination of different conservation and preservation strategies. It is a philosophy which sees humans become 'caretakers' of the natural world. There is overlap here with the concept of sustainable development: as stewards, we care for the physical environment so that it can be inherited by successive generations, thereby delivering inter-generational equity.

Balanced stewardship approaches to resources management can be illustrated by looking at efforts to manage and use marine (ocean) ecosystems more sustainably. Table 3.10 shows a range of stewardship actions carried out at different scales by a variety of different stakeholders.

3.3 Resource stewardship possibilities

Table 3.10 Stewardship actions in relation to the way ocean resources are managed and consumed

Global actions	• The **United Nations Food and Agriculture Organization (FAO)** aims to 'ensure long-term conservation and sustainable use of marine living resources in the deep sea and to prevent significant adverse impacts on vulnerable marine ecosystems'. As part of its stewardship work, the FAO can designate marine protected areas (MPAs) in the high seas. However, many illegal fishing activities still occur due to a lack of any real means of enforcement. • World Oceans Day is a 'global day of ocean celebration and collaboration for a better future' held every year on 8 June. This collaboration between the charity The Ocean Project, the United Nations and many other partners raises awareness of ocean issues.
International and national actions (fishing quotas/limits)	• The **EU Common Fisheries Policy (CFP)** is a set of stewardship rules designed to conserve fish stocks in European waters. Catch limits called total allowable catches (TACs) are regularly updated for commercial fish stocks using the latest scientific advice on fish stock status. TACs are shared between EU countries in the form of national quotas. • Sometimes these quotas are controversial because they affect livelihoods for people who fish.
Local actions (no-catch and conservation zones)	• Much of the waters of western Scotland have been overfished. The community living on the island of Arran established an organization called COAST (Community of Arran Seabed Trust). They successfully lobbied the Scottish government to designate local waters as Scotland's first **no-take zone**. • All fishing within the specified area has been banned. But as a result, some fishermen have become unemployed.
Businesses (aquaculture production)	• Globally, aquaculture has grown rapidly since the 1980s. Salmon, cod, trout, scallops and prawns are now reared in huge volumes in cages in coastal waters. • However, fish farming can lead to the outbreak of diseases that spread to wild populations.
Citizens (campaigning and consumption)	• Individual consumers choose not to buy threatened species, or to eat fewer fish. • Civil society organizations (such as the UK's 'Fish Fight' campaign) try to educate the public about sustainability issues in the hope that behaviours will change and more people will adopt stewardship principles.

> **PPPPSS CONCEPTS**
>
> Think about the different stakeholders and actions shown in Table 3.10. In your view, who has the greatest power to make a real difference in terms of how ocean resources are managed for the future? Why?

> **Keyword definitions**
>
> **No-take zone** An area of water where fishing has been banned completely.
>
> **Sustainable development** Meeting the needs of the present without compromising the ability of future generations to meet their own needs.

Progress towards sustainable development

The stewardship strategies shown in Table 3.10 are part of a greater quest for **sustainable development**. Widely adopted after the 1992 UN Conference on Environment and Development in Rio, sustainable development means: 'meeting the needs of the present without compromising the ability of future generations to meet their own needs'.

Three goals form the basis of sustainability, or sustainable development (Figure 3.31).
- **Economic** sustainability: individuals and communities should have access to a reliable income over time.
- **Social** sustainability: all individuals should enjoy a reasonable quality of life.
- **Environmental** sustainability: no lasting damage should be done to the environment; renewable oceanic, terrestrial and atmospheric resources must be managed in ways that guarantee continued use.

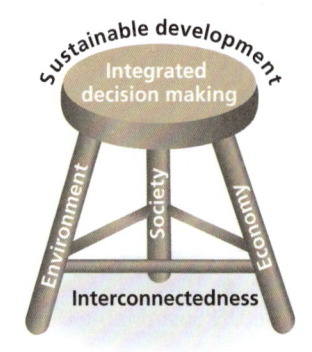

Figure 3.31 A model of sustainable development

For the last of these goals to be met, there must be either a significant reduction in world economic output, or new technological fixes that increase resource availability and repair environmental damage. The former may not happen voluntarily; this is because economic growth is the goal of free market economies. The latter may prove expensive to implement on a scale large enough to be effective. Actions intended to support sustainable management of the oceans should, therefore, be assessed with a critical eye.

Units 1 and 2 have demonstrated that current patterns of consumption across the world are no longer sustainable; the ability of Earth to meet our water, food, energy and other needs now and in the future is seriously jeopardized. Much needs to be done and two important steps in the right direction are the UN's Sustainable Development Goals initiative and the growing commercial adoption of **circular economy** approaches to resource use and waste management.

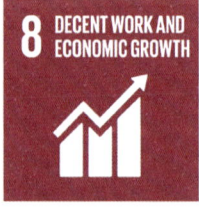

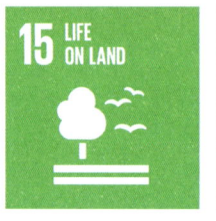

The UN Sustainable Development Goals

The Sustainable Development Goals (SDGs) are a revised and enlarged version of the earlier Millennium Development Goals (MDGs) of 2000–15 (which committed world leaders to combat poverty, hunger, disease, illiteracy, environmental degradation and discrimination against women). The post-2015 transformative plan of action is now based around 17 interrelated goals. Collectively, these address what experts view as the most urgent contemporary global challenges. Most of them feature prominently in this book and they are shown in Figure 3.32.

- Both the SDGs and earlier MDGs provide a 'roadmap' for human development by setting out priorities for action.
- The SDGs integrate and balance the three economic, social and environmental dimensions of sustainable development shown in Figure 3.32 as part of a comprehensive global vision.

> **PPPPSS CONCEPTS**
> To what extent do the strategies outlined in Table 3.10 operate on a sufficiently large or bold scale to have a significant positive impact on global sustainable development?

> **Keyword definition**
> **Circular economy** An approach to business management and product design that maximizes the efficiency of resource use, and aims ultimately to phase out waste and pollution altogether.

> **PPPPSS CONCEPTS**
> Think about the possibility of all or any of these goals being achieved by 2030. Use evidence and arguments from this book to support your views.

Figure 3.32 The UN Sustainable Development Goals for 2030

At the time of writing, the UN's latest SDG progress report was available online at **https://unstats.un.org/sdgs/report/2016/**. The document includes several headline findings that are relevant to your course and which reinforce important points this book has made previously:
- **Goal 1 No poverty** 'The proportion of the global population living below the extreme poverty line dropped by half between 2002 and 2012, from 26 to 13 per cent.' (This success is analysed in this book on pages 77–80.)
- **Goal 2 Zero hunger** 'Nearly 800 million people worldwide still lack access to adequate food.' (This food security challenge is addressed on pages 86–87.)
- **Goal 5 Gender equality** 'There were still 757 million adults (aged 15 and over) unable to read and write, of whom two-thirds were women.' (Gender inequality is explored on pages 29–32.)
- **Goal 6 Clean water and sanitation** '6.6 billion people, or 91 per cent of the global population, used an improved drinking water source, compared with 82 per cent in 2000. But water stress affects more than 2 billion people around the globe, a figure that is projected to rise.' (Water security is the focus of pages 84–85.)
- **Goal 7 Affordable and clean energy** 'Some 3 billion people, over 40 per cent of the world's population, relied on polluting and unhealthy fuels for cooking. Modern renewables grew rapidly, at a rate of 4 per cent a year between 2010 and 2012.' (Global patterns of energy use are described on pages 88–90.)
- **Goal 13 Climate action** 'The historic Paris Agreement sets the stage for ambitious climate action by all to ensure that global temperatures rise no more than 2°C.' (This success is reported on pages 67–68.)
- **Goal 14 Life below water** 'The proportion of global marine fish stocks within biologically sustainable levels declined from 90 per cent in 1974 to 69 per cent in 2013.' (The response to this stewardship challenge appears in Table 3.10 on page 109.)

> **PPPPSS CONCEPTS**
>
> Think about how greater progress towards some of the goals highlighted here might possibly threaten progress towards other goals.

The circular economy

The focus of this book's final section is a resource stewardship approach called the **circular economy**. This is a way of thinking about resource use which aims to jointly nurture ecological and economic health in order to promote truly sustainable development. It is a far bolder vision than either increased recycling or improved efficiency of primary resource and fossil fuel use (which merely delays the inevitable exhaustion of finite reserves). Its philosophy derives from the study of natural systems (such as ecosystems, or the water and carbon cycles).

Traditional, linear systems often use natural resources (including water, food and oil) wastefully to create products. When goods are sold, ownership passes from producer to consumer. This includes responsibility for waste disposal when an item is used up or breaks – and as a result, most waste ends up in landfill sites (Figure 3.33).

In contrast, the circular economy has the following characteristics:
- All outputs are reprocessed and all waste is viewed as a resource. Reuse may be more efficient than recycling. Vehicles, plastics and buildings are all designed with a view to dismantling and reusing parts later: this is called 'closing the loop'.
- Food is managed more carefully at all stages of the supply chain. All waste is composted. Natural resource stocks are maintained, benefiting both the environment and society.
- To optimize the function of a circular economy, a performance economy can be operated as part of the equation: people rent or share goods instead of buying them. This way, the manufacturer retains ownership of the product (and its embodied resources) and responsibility for waste. Companies are encouraged to innovate more creative and cost-effective ways of recycling and reusing.

Figure 3.34 portrays the circular economy as twin biological and technical cycles operating in tandem. This cradle-to-cradle framework was designed by architect William McDonough and chemist Michael Braungart. The key aim is to progress beyond current linear and wasteful economic systems and to ultimately 'design out' waste altogether.

Since 2010, the Ellen MacArthur Foundation has actively promoted the circular economy concept and in 2015 the European Commission submitted a Circular Economy Package to the European Parliament. Circular economy concepts have been successfully applied at a small scale since the 1990s (for instance, in 'industrial ecosystem' parks such as the Kalundborg Symbiosis in Denmark).

Figure 3.33 Will landfill become a thing of the past if circular economy approaches are adopted?

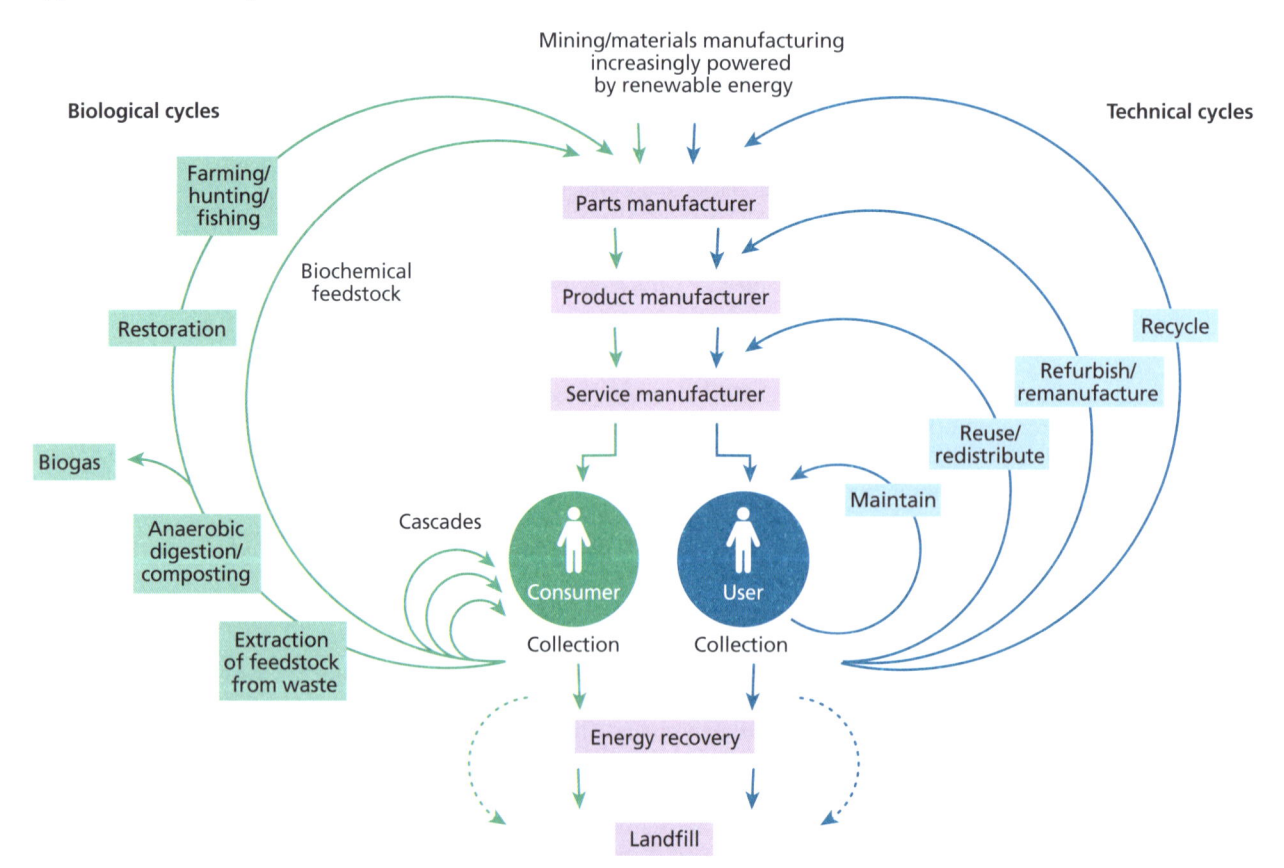

Figure 3.34 The twin biological and technical loops of the circular economy

Important steps towards the implementation of circular economy principles have already been taken at varying geographic scales (Table 3.11). Notably, the 2015 Sustainable Development Goals (SDGs) include a commitment to reduce waste in food systems globally (Goal 12).

Table 3.11 Examples of the application of circular economy principles at varying scales

Global governance	The SDG target is to halve food waste worldwide by 2030. The EU has committed to this target.
	The EU already has long-standing 'circular' legislation in place: the 2003 Directive on waste electrical and electronic equipment (WEEE Directive) provided for the creation of collection schemes where consumers return their WEEE, such as fridges, to the manufacturer free of charge.
National targets	The US Department of Agriculture and Environmental Protection Agency has set a national food waste reduction goal (matching the UN's global goal).
	France and Italy have taken steps to make it illegal for supermarkets to throw out unsold food, requiring instead that they donate it to charities, or animal feed and composting companies.
Local initiatives	Suzhou New District, near Shanghai in China, is home to 4,000 manufacturing firms, many of whom have interlinked their operations and become interdependent. Manufacturers of printed circuit boards use copper that is recovered from waste generated by other companies in the park (rather than using virgin copper produced by mining firms elsewhere).

■ Can the circular economy succeed?

A lack of commercial incentive is preventing wider uptake of circular thinking currently. To achieve circular economy goals, companies everywhere must design products that last longer and/or make greater use of recycled materials. Yet if costs rise and profits fall, where is the incentive? The current low oil prices provide a case in point. Owing to the falling cost of the key raw material for plastic (oil), the production of new plastic has become less expensive than the process required to recycle old plastic into new materials. This is because the cleaning and preparation of used plastics takes a lot of water, energy and effort (and there are worrying WFE nexus implications here too). The circular economy may therefore not succeed if **neoliberal** beliefs prevent governments from intervening strategically in markets. It may be a long time before oil prices fall, for instance. If that is the case, then governments will need to take action (instead of assuming that businesses and civil society will adopt circular economy principles voluntarily).

Pessimists might therefore view circular economy thinking as 'too little, too late'. They may point despairingly towards the vast quantities of resources and waste associated with China's production of half the world's aluminium, steel and cement. Yet China increasingly leads the way in promoting renewable energy and in setting targets for reusing waste. Unlike successive neoliberal governments of the USA, Chinese leaders are often quick to impose legislation to accelerate changes.

In contrast, optimists are excited by recent socio-technical changes that both support and are shaped by circular thinking. Longer-lasting products such as graphene are being used already in manufacturing; new edible forms of plastic are being developed. The fast-maturing technology of driverless cars could play a pivotal role in making product rental – and not ownership – the new social norm. Finally, sociologists have observed that many Millennials participate willingly in a 'sharing economy' and are committed to sustainable development goals. This generation is now part of the political decision-making process.

Optimists may therefore conclude that the technological, economic and – perhaps more importantly – political building blocks for a circular economy are already in place.

> **Keyword definition**
>
> **Neoliberal** A philosophy of managing economies and societies which takes the view that government interference should be kept to a minimum and that problems are best left for market forces to solve.

■ KNOWLEDGE CHECKLIST:

- Pessimistic views on population growth, including Malthus and the Neo-Malthusians (Club of Rome)
- Optimistic views on population growth, including the Boserup thesis
- Balanced views on the population–resources relationship, including the concepts of resource stewardship, resource conservation and resource preservation
- The power of different stakeholders to promote stewardship approaches (in relation to ocean resource management)
- The concept of sustainable development and the possibility of achieving its aims
- The role of the Sustainable Development Goals (SDGs) as a human development and resource stewardship 'roadmap'
- Progress made so far towards SDG targets being met
- The value of the circular economy approach for sustainable development

EVALUATION, SYNTHESIS AND SKILLS (ESK) SUMMARY:

- How perspectives differ on the possibility of avoiding positive checks to population growth
- How different management approaches have been adopted at varying geographic scales in order to achieve more sustainable outcomes for people and the planet.

114 Unit 3 Global resource consumption and security

EXAM FOCUS

INFOGRAPHIC ANALYSIS

Section B of your examination consists of an 'infographic analysis'. In Section A, you are required to describe data, such as a graph showing a trend, or the pattern on a map. The task in Section B takes 'graphical literacy' one step further by requiring that you additionally *evaluate* the strengths and weaknesses of the way data are presented. You will be shown an infographic and, among other things, may be asked to:

- explain strengths and weaknesses in the presentation methods used (an AO2 task)
- suggest possible improvements that could be made (an AO2 task)
- evaluate the infographic for evidence of possible bias in the way facts are presented or language has been used (an AO3 task).

Figure 3.35 shows an exam-style infographic. It is followed by a series of questions, each of which is accompanied by a student answer and examiner commentary.

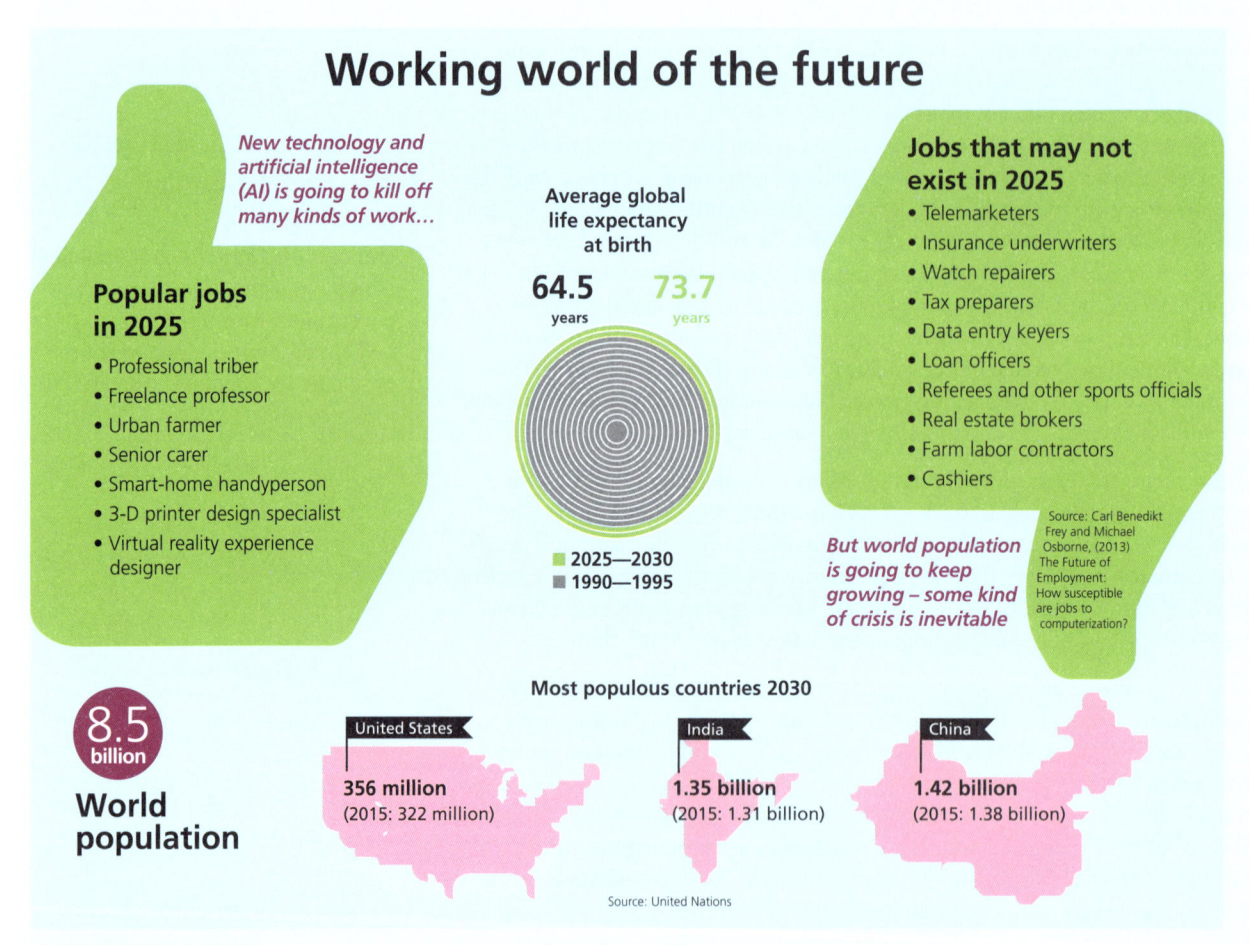

Working world of the future

New technology and artificial intelligence (AI) is going to kill off many kinds of work…

Popular jobs in 2025
- Professional triber
- Freelance professor
- Urban farmer
- Senior carer
- Smart-home handyperson
- 3-D printer design specialist
- Virtual reality experience designer

Average global life expectancy at birth

64.5 years   73.7 years

■ 2025—2030
■ 1990—1995

Jobs that may not exist in 2025
- Telemarketers
- Insurance underwriters
- Watch repairers
- Tax preparers
- Data entry keyers
- Loan officers
- Referees and other sports officials
- Real estate brokers
- Farm labor contractors
- Cashiers

Source: Carl Benedikt Frey and Michael Osborne, (2013) The Future of Employment: How susceptible are jobs to computerization?

But world population is going to keep growing – some kind of crisis is inevitable

8.5 billion World population

Most populous countries 2030

United States: 356 million (2015: 322 million)

India: 1.35 billion (2015: 1.31 billion)

China: 1.42 billion (2015: 1.38 billion)

Source: United Nations

Source: Adapted from http://taxinsights.ey.com/system-admin/editions/tax15-talent-and-tax-abo-version-lowres-160316-4a.pdf

Figure 3.35 A look at the working world of the future

a **Identify which country has experienced:**
   i **the smallest increase in the number of people**
   ii **the greatest proportional increase in population size. [1 + 1 marks]**

i The USA
ii India

Examiner comment

Both answers are correct. The first question requires you to compare millions with billions, which might lead some to give the wrong answer. The second question requires you to estimate proportional increases. India – which has increased from 1,310 million to 1,530 million – has grown proportionately larger than the USA, which has risen from 322 million to 356 million.

b Explain TWO weaknesses in the way this infographic presents its facts. [2 + 2 marks]

1 Some of the language used is not easy to understand and relies on the reader having specialist knowledge. It is not clear what a 'professional triber' is. A good infographic might include footnotes to help explain a term like this.

2 The graphic used to show the increase in life expectancy is unhelpful. It is not very clear what the circles are meant to represent.

Examiner comment

Overall this answer might score 3 out of the 4 possible marks. The first point is clear and well-explained. The weakness is made very clear when the student explains what else might be expected of a good infographic (i.e. footnotes). The second answer is less clear. It is true that the circles used to show the increase in average global life expectancy do not convey information in an easy-to-understand way. However, for 4 marks we might expect the answer to go a bit further, perhaps by identifying the infographic's failure to use actual proportional circles to display the information.

c To what extent does this infographic promote a *pessimistic* view about population growth? [6 marks]

On the one hand, the infographic paints a very negative picture of the future. It does this by using very strong language, such as the word 'kill'. The phrase 'some kind of crisis is inevitable' sounds very sensationalist too. It is also an opinion because not everyone agrees that a Malthusian crisis will occur this century.

On the other hand it is not entirely pessimistic. New jobs are shown, which suggests that new technologies are developing as time passes, much as Boserup argued. The 'urban farmers' might be using new food technology developed to cope with growing populations. The increase in life expectancy that is shown is good news, because it shows that UN sustainable development goals are being reached and more people are living to reach an old age.

Overall, however, the tone is more negative and pessimistic than positive.

Examiner comment

This is a well-structured answer that shows clear understanding of what is meant by the command phrase 'To what extent...' (there is an expectation that an answer should argue both *for and against* the viewpoint that is being put forward, i.e. that the infographic is pessimistic). The response is also grounded in knowledge and understanding of geographical theory, including the writing of Malthus and Boserup. This means that a developed answer is provided which is both knowledgeable and authoritative.

A number of pessimistic and optimistic views are identified and explained before a brief final judgement is arrived at. This gives a good indication of the kind of answer that might be expected from a student if full marks were to be gained using no more than the 8–9 minutes of reading and writing time allowed for this type of question in the actual examination (overall, the exam lasts 75 minutes and there are 50 marks to be gained in total).

Glossary

Adaptation – Any action designed to protect people from the harmful impacts of climate change but without tackling the underlying problem of rising GHG emissions.

Ageing population – A population structure where the proportion of people aged 65 and over is high and rising. This is caused by increasing life expectancy and can be further exaggerated by the effect of low birth rates. It is also called a 'greying' population.

Albedo – How much solar radiation a surface reflects. White surfaces have the highest albedo, or reflectivity.

Anthropogenic Carbon flow – The current amount of carbon emissions released annually by a country (e.g. due to fossil fuel burning and cement making) produced in each nation. The figure can be adjusted upwards to factor in the carbon equivalents of other greenhouse gas emissions (methane and nitrous oxide).

Anthropogenic Carbon stock – The total size of the store of anthropogenic (human) carbon emissions released into the atmosphere since industrialization began around 1750.

Arid – A climate whose precipitation is less than 250 mm annually.

Backwash – Flows of people, investment and resources directed from peripheral to core regions. This process is responsible for the polarization of regional prosperity between regions within the same country.

Biome – Large planetary-scale plant and animal community covering large areas of the Earth's continents. For example, tropical rainforest, desert and grassland.

Cap and trade – An environmental policy that places a limit on the amount a natural resource can be used, specifies legitimate resource users, divides this amount up into shares per user, and allows users to sell their shares if they do not wish to use them directly.

Carbon intensity – The amount of CO_2 emitted per unit of GDP. If a country's carbon emissions rise less slowly than its GDP is increasing, this suggests some action is being taken to reduce emissions at the same time as industrial output is increasing.

Carrying capacity – The maximum number of people an area of land can support with current levels of technology.

Circular economy – An approach to business management and product design that maximizes the efficiency of resource use, and aims ultimately to phase out waste and pollution altogether.

Civil society – Any organization or movement that works in the area between the household, the private sector and the state to negotiate matters of public concern. Civil society includes non-governmental organizations (NGOs), community groups, trade unions, academic institutions and faith-based organizations.

Climate change – Any long-term trend or movement in climate detected by a sustained shift in the average value for any climatic element (e.g. rainfall, drought, hurricanes).

Consumption – The level of use a society makes of the resources available to it. Economic development and changing lifestyles and aspirations usually result in accelerated consumption of resources.

Core–periphery system – The uneven spatial distribution of national population and wealth between two or more regions of a state or country, resulting from flows of migrants, trade and investment.

Cryosphere – Those portions of Earth's surface where water is in solid form.

Deindustrialization – The loss of traditional manufacturing industries in some high-income countries due to their closure or relocation elsewhere. Since the 1960s, many industries have all but vanished from Europe and North America. Instead, they thrive in Asia, South America and, increasingly, Africa.

Demography – The study of population dynamics and changes.

Desalinization – The removal of salt water and other minerals from seawater. The process is costly and requires desalinization plants to be built.

Desertification – The intensification or extensification of arid, desert-like conditions.

Development – Human development generally means the ways in which a country seeks to progress economically and to also improve the quality of life for its inhabitants. A country's level of development is shown firstly by economic indicators of average national wealth and/or income, but can encompass social and political criteria too.

Ecological footprint – A crude measurement of the area of land or water required to provide a person (or society) with the energy, food and resources needed to live, and to also absorb waste.

El Niño Southern Oscillation (ENSO) – A sustained sea surface temperature anomaly across the central tropical Pacific Ocean. It brings a change in weather conditions that can last 2–7 years. Along with La Niña events, El Niño events are part of a short-term climate cycle that brings variations in climate but only for a few years.

Embedded water – A measure of the amount of water used in the production and transport to market of food and commodities (also known as the amount of 'virtual water' or 'water footprint' attached to a product). Embedded water may include the use of local water resources and the use of water resources in distant places.

Energy mix – The proportions of hydrocarbons, renewable energy sources and nuclear energy that a country uses to meet its domestic needs.

Energy pathways – Flows of energy that link producer regions with consumer regions. In physical terms, pathways take the form of pipelines, electricity power lines and the routes taken by gas and oil tankers, or trains carrying coal.

Acknowledgements

The Publishers would like to thank the following for permission to reproduce copyright material.

Photo credits

p.19 © Didier Marti/123RF; **p.20** © Shariful Islam/Xinhua/Alamy Stock Photo; **p.22** © Cmacauley/https://commons.wikimedia.org/wiki/File:Arbat_Transit_Camp_3-3-2014.jpg; **p.27** © 2005 Warner Bros. Entertainment Inc. - Harry Potter Publishing Rights © J.K.R.; **p.40** © Ashley Cooper/Global Warming Images/Alamy Stock Photo; **p.50** © NASA; **p.51** © Anton Goida/123RF; **p.55** © NASA; **p.56** © Peter Prokosch/GRID-Arendal; **p.64** *t* © Planetpix/NASA Photo/Alamy Stock Photo, *b* © Eoghan Rice/Trocaire/https://commons.wikimedia.org/wiki/File:Tacloban_Typhoon_Haiyan_2013-11-14.jpg; **p.65** © Sung kuk kim/123RF; **p.66** © Horizon magazine; **p.73** © f11photo/Shutterstock; **p.101** © Marlenenapoli/https://commons.wikimedia.org/wiki/File:Agbogbloshie.JPG; **p.112** © Cezary p/https://commons.wikimedia.org/wiki/File:Wysypisko.jpg

t = top, *b* = bottom, *l* = left, *r* = right, c = centre

Text credits

p.17 Figure 1.17 from http://www.ons.gov.uk/ons/rel/pop-estimate/population-estimates-for-uk--england-and-wales--scotland-and-northern-ireland/mid-2011-and-mid-2012/sty---uk-population-estimates.html. Contains public sector information licensed under the Open Government Licence v3.0 (http://www.nationalarchives.gov.uk/doc/open-government-licence/version/3/); **p.18** Figure 1.19 from United Nations, Department of Economic and Social Affairs, Population Division (2011): *World Population Prospects: The 2010 Revision.* New York; **p.29** Figure 1.29 from OECD (2012), Gender, Institutions and Development Database 2012 (GID-DB) http://stats.oecd.org/Index.aspx?datasetcode=GIDDB2012 (accessed on 24 April 2017); **p.44** Figure 2.10 data taken from http://www.britishgas.co.uk/the-source/carbon-emissions © British Gas; **pp.58–59** Figure 2.24 adapted from http://www.metoffice.gov.uk/climate-guide/climate-change/impacts/four-degree-rise/map. Contains public sector information licensed under the Open Government Licence v3.0 (http://www.nationalarchives.gov.uk/doc/open-government-licence/version/3/); **p.65** Figure 2.29 taken from https://www.gov.uk/government/organisations/environment-agency. Contains public sector information licensed under the Open Government Licence v3.0 (http://www.nationalarchives.gov.uk/doc/open-government-licence/version/3/); **p.77** Figure 3.2 data taken from Povcalnet World Bank & Ferreira et al. (World Bank, 2015); graph CC BY-SA 4.0 (https://creativecommons.org/licenses/by-sa/4.0/deed.en_US); **p.80** Figure 3.5 data taken from Lakner and Milanovic (2015) 'Global Income Distribution: From the Fall of the Berlin Wall to the Great Recession', *World Bank Economic Review, Advance Access* published August 12, 2015; graph CC BY-SA by the authors Zdenek Hyne; **p.83** Figure 3.7 taken from *China Ecological Footprint Report 2012* © WWF/Global Footprint Network (http://www.footprintnetwork.org/content/images/article_uploads/China_Ecological_Footprint_2012.pdf); **p.85** Figure 3.9 taken from Chapagain, A. K. and S. Orr (2008) *UK Water Footprint: The impact of the UK's food and fibre consumption on global water resources, Volume 1*, WWF-UK, Godalming, UK; **p.88** Figure 3.12 taken from *BP Statistical Review of World Energy 2016* (http://www.bp.com/content/dam/bp/pdf/energy-economics/statistical-review-2016/bp-statistical-review-of-world-energy-2016-full-report.pdf); **p.93** Figure 3.17 We are grateful to Chatham House, the Royal Institute of International Affairs, for permission to reproduce the work entitled: 'Figure C: Share of global production (>5%) of key commodities and water scarcity' originally published in Bernice Lee, Felix Preston, Jaakko Kooroshy, Rob Bailey and Glada Lahn, 2012, *Resources Future*; **p.94** Figure 3.18 Food and Agriculture Organization of the United Nations, 2004, Olivier Dubois, Jean-Marc Faurès, Erika Felix, Alessandro Flammini, Jippe Hoogeveen, Lucie Pluschke, Manas Puri and Olçay Ünver, *The Water-Energy-Food Nexus: A new approach in support of food security and sustainable agriculture.* Reproduced with permission; **p.110** Figure 3.32 from http://www.un.org/sustainabledevelopment/sustainable-development-goals/ © United Nations 2017. Reprinted with the permission of the United Nations; **p.114** Figure 3.35 from *Tax Insights for Business Leaders No 15* © EYGM Limited (http://taxinsights.ey.com/system-admin/editions/tax15-talent-and-tax-abo-version-lowres-160316-4a.pdf).

Every effort has been made to trace all copyright holders, but if any have been inadvertently overlooked, the Publishers will be pleased to make the necessary arrangements at the first opportunity.